AF581268

Monographs in Mathematics
Vol. 98

Victor Palamodov

Reconstructive Integral Geometry

Birkhäuser Verlag
Basel · Boston · Berlin

Author:
Victor Palamodov
School of Mathematics
Tel Aviv University
Ramat Aviv
Tel Aviv 69978
Israel
e-mail: palamodo@tau.ac.il

2000 Mathematics Subject Classification: primary 44A12, 53C65, 65R32; secondary 35C15, 35N10, 92C55

A CIP catalogue record for this book is available from the
Library of Congress, Washington D.C., USA

Bibliographic information published by Die Deutsche Bibliothek
Die Deutsche Bibliothek lists this publication in the Deutsche Nationalbibliografie; detailed bibliographic data is available in the Internet at <http://dnb.ddb.de>

ISBN 3-7643-7129-3 Birkhäuser Verlag, Basel – Boston – Berlin

Part of Springer Science+Business Media
Printed on acid-free paper produced from chlorine-free pulp. TCF ?
Printed in Germany
ISBN 3-7643-7129-3

9 8 7 6 5 4 3 2 1

To my family

Contents

Preface

One hundred years ago (1904) Hermann Minkowski [58] posed a problem: to reconstruct an even function f on the sphere S^2 from knowledge of the integrals

$$Mf\,(C) = \int_C f\mathrm{d}s$$

over big circles C. Paul Funk found an explicit reconstruction formula for f from data of big circle integrals. Johann Radon studied a similar problem for the Euclidean plane and space. The interest in reconstruction problems like Minkowski-Funk's and Radon's has grown tremendously in the last four decades, stimulated by the spectrum of new modalities of image reconstruction. These are X-ray, MRI, gamma and positron radiography, ultrasound, seismic tomography, electron microscopy, synthetic radar imaging and others. The physical principles of these methods are very different, however their mathematical models and solution methods have very much in common. The umbrella name *reconstructive integral geometry*[1] is used to specify the variety of these problems and methods.

The objective of this book is to present in a uniform way the scope of well-known and recent results and methods in the reconstructive integral geometry. We do not touch here the problems arising in adaptation of analytic methods to numerical reconstruction algorithms. We refer to the books [61], [62] which are focused on these problems.

Various aspects of interplay of integral geometry and differential equations are discussed in Chapters 7 and 8. The results presented here are partially new.

The book is an extended version of a lecture course which was read for students of Tel Aviv University. Chapter 1 contains a mini-course in distribution theory, harmonic analysis and geometry. Not much of this knowledge is necessary for reading and understanding Chapters 2, 3 and 6.

[1]Blaschke's term "Integralgeometrie" looks somehow redundant; the word geometry itself has, since ancient Greek times, meant calculation of lengths, areas, volumes, i.e., some integrals.

Notation

$\mathbb{Z}$ – the algebra of integers,
$\mathbb{R}$ – the field of real numbers,
$\mathbb{R}_+$ – the set of non-negative numbers
$\mathbb{C}$ – the field of complex numbers
$\mathrm{j} \doteq 2\pi\imath,\ \imath \doteq \sqrt{-1}$
$f_+ = \max(f, 0),\ f_- = \max(-f, 0)$
$\Gamma(\lambda),\ \lambda \in \mathbb{C}$ – Gamma-function
$\mathbf{V}$ – a vector space over $\mathbb{R}$ of finite dimension
$\mathbf{V}^*$ – the space dual to a vector space $\mathbf{V}$
$\mathrm{d}x$ or $\mathrm{d}V$ – the volume form in a coordinate space $\mathbf{V}$
$\mathbf{E}$ or $\mathbf{E}^n$ – a Euclidean space of dimension $n < \infty$
$D(\mathbf{V})$ – the space of smooth functions of compact support (test functions) in the vector space $\mathbf{V}$
$K(\mathbf{V})$ – the space of smooth densities of compact support (test densities) in $\mathbf{V}$
$S(\mathbf{V})$ – the space of smooth fast decreasing functions in $\mathbf{V}$
$F(f) = \hat{f}$ – the Fourier transform of a function f or of the density $f\,\mathrm{d}x$ defined in $\mathbf{V}$; $\hat{f}(\xi) = \int_{\mathbf{V}} \exp(-\mathrm{j}\xi x) f(x)\,\mathrm{d}x$
$F^*(f) = \int_{\mathbf{V}^*} \exp(\mathrm{j}\xi x) f(\xi)\,\mathrm{d}\xi$ – the adjoint Fourier transform

Chapter 1

Distributions and Fourier Transform

1.1 Introduction

The Riemann theory of integration in a real vector space of finite dimension is sufficient for our purposes except for Section 1.7, where we outline the Fourier-Plancherel theory. Let $\mathbf{V}$ be a real vector space of finite dimension n. Choose a coordinate system $x = (x_1, \ldots, x_n)$ and take the measure $\mathrm{d}x = \mathrm{d}x_1 \ldots \mathrm{d}x_n$ in $\mathbf{V}$. A function $f : \mathbf{V} \to \mathbb{C}$ is called *Riemann integrable*, if it is continuous almost everywhere in $\mathbf{V}$, (i.e., except for a zero measure set) and the norm

$$\|f\|_1 \doteq \sup_{B,t} \int_B |f_t| \, \mathrm{d}x$$

is finite, where B denotes an arbitrary ball in $\mathbf{V}$ and $f_t = f$ if $|f| \leq t$ and $f_t = 0$ otherwise (truncation of f). We set

$$\int_{\mathbf{V}} f \, \mathrm{d}x \doteq \lim_{r,t \to \infty} \int_{B(r)} f_t \mathrm{d}x$$

where $B(r)$ is the ball of radius r centred at the origin (cubes can be taken instead of balls etc.). We shall use standard tools of Riemann integration theory including Fubini's Theorem. The full range Fubini theorem will be stated in Section 1.7 in the framework of Lebesgue integration theory.

1.2 Distributions and generalized functions

Let $\mathbf{V}$ be a finite-dimensional vector space, U an open set in $\mathbf{V}$. A function $f : U \to \mathbf{C}$ is called *smooth* or of class C^∞, if it has all derivatives with respect to a system of coordinates in $\mathbf{V}$. This property is equivalent to the condition: f has a differential of arbitrary order at each point $x \in \mathbf{V}$ and is preserved under any linear (or smooth) coordinate change.

Definition. The notation $D = D(U)$ means the space of smooth functions ϕ in $\mathbf{V}$ with compact support $\operatorname{supp} \phi \Subset U$ (called test functions). This is a vector space over the field $\mathbb{C}$ of complex numbers. The natural convergence in $D(U)$ is defined as follows: $\{\phi_k, k = 1, 2, \ldots\}$ converges to ϕ, if

(i) $\cup \operatorname{supp} \phi_k \subset K \Subset U$ and
(ii) $D^q(\phi_k - \phi) \rightrightarrows 0$ uniformly in U for arbitrary $q = (q_1, \ldots, q_n)$, where

$$D^q = \frac{\partial^{q_1+\cdots+q_n}}{(\partial x_1)^{q_1} \ldots (\partial x_n)^{q_n}}$$

and $x_1, \ldots, x_n$ is a linear coordinate system in $\mathbf{V}$. A functional $u : D(U) \to \mathbb{C}$ is called continuous with respect to the natural convergence, if $u(\phi_k) \to u(\phi)$ for any convergent sequence $\phi_k \to \phi$. Any linear continuous functional on $D(\mathbf{V})$ is called Schwartz's *distribution* in U. The space of all distributions is denoted $D'(U)$.

A test density in U is a smooth density ρ with compact support. For a choice of the coordinate system we can write $\rho = \phi \mathrm{d}x$ where $\mathrm{d}x = \mathrm{d}x_1 \wedge \cdots \wedge \mathrm{d}x_n$ and $\phi = \rho/\mathrm{d}x$. The space of test densities with support in U is denoted $K(U)$. This space is supplied with the natural convergence: a sequence ρ_k converges to ρ as the sequence of test functions $\phi_k = \rho_k/\mathrm{d}x$ converges to $\phi \doteq \rho/\mathrm{d}x$ in the sense (i)–(ii). A linear continuous functional on the space $K(U)$ is called Sobolev's *generalized function* in U.[1] The linear operations, multiplication by a smooth function and derivatives are well defined in the space $D'(U)$ of distributions and in the space $K'(U)$ of generalized functions.

Example 1. A function $f : V \to \mathbb{C}$ is called locally integrable, if its restriction to an arbitrary compact set $K \subset U$ is integrable. Any locally integrable function f defines a generalized function $[f]$ as

$$[f](\rho) = \int f \rho = \int f \phi \mathrm{d}x, \; \rho = \phi \mathrm{d}x \in K(U).$$

The functional $[f]$ equals zero if and only if the function f vanishes almost everywhere. The density $f \mathrm{d}x$ defines a distribution denoted $[f \mathrm{d}x]$.

Example 2. A Dirac function at a point $a \in \mathbf{V}$ is the functional $\delta_a(\phi \mathrm{d}x) = \varphi(a)$ on the space $K(\mathbf{V})$. Dirac-distribution $\delta_a \mathrm{d}x$ is the continuous functional on $D(\mathbf{V})$: $\delta_a \mathrm{d}x(\phi) = \phi(a)$.

Example 3. For a function $\phi : \mathbb{R} \to \mathbb{C}$ the limit

$$[\frac{\mathrm{d}x}{x}](\phi) \doteq \lim_{\varepsilon \searrow 0} \int_{|x|>\varepsilon} \frac{\phi \mathrm{d}x}{x}$$

is called the *principal value* of the divergent integral. The limit exists for any integrable function ϕ with compact support that has derivative at the point $x = 0$. The functional $[\mathrm{d}x/x]$ is linear and continuous in $D(\mathbb{R})$, i.e., a distribution in $\mathbb{R}$.

---

[1] de Rham's construction of *currents* [18] contains generalized functions (currents of degree 0) and distributions (currents of degree n) as particular cases.

Example 4. EULER KERNELS. For an arbitrary complex λ, $\operatorname{Re}\lambda > 0$, we define the functional on $S(\mathbb{R})$,

$$H^{\lambda}(\phi) = \frac{x_{+}^{\lambda-1}\mathrm{d}x}{\Gamma(\lambda)}(\phi) \doteq \frac{1}{\Gamma(\lambda)}\int_0^{\infty} x^{\lambda-1}\phi(x)\,\mathrm{d}x,$$

where $x^{\lambda-1} \doteq \exp((\lambda-1)\ln x)$, $\ln x \in \mathbb{R}$ for $x > 0$. This is a tempered distribution which depends analytically on λ, i.e., $H^{\lambda}(\phi)$ is a holomorphic function of λ for each ϕ.

Proposition 1.1 *The family H^{λ} has holomorphic continuation on the whole complex plane $\mathbb{C}$. We have $H^{-k} = \delta^{(k)}(0)$ for $k = 0, 1, 2, \ldots$*

◀ We have for $\operatorname{Re}\lambda > 1$, for $\phi' = \mathrm{d}\phi/\mathrm{d}x$,

$$\left(H^{\lambda}(\phi)\right)' = -H^{\lambda}(\phi') = -\frac{1}{\Gamma(\lambda)}\int x^{\lambda-1}\phi'(x)\,\mathrm{d}x = \frac{1}{\Gamma(\lambda-1)}\int x^{\lambda-2}\phi(x)\,\mathrm{d}x,$$

where we integrate by parts and take into account that $x^{\lambda-1}\phi(x)$ vanishes at $x = 0$ and at $x = \infty$. The right side is equal to $H^{\lambda-1}(\phi)$ and the differential equation $\left(H^{\lambda}\right)' = H^{\lambda-1}$ holds. Now we can define $H^{\lambda} \doteq \left(H^{\lambda+1}\right)'$ for $\operatorname{Re}\lambda > -1$. This formula defines analytic family H^{λ} which coincides with the Euler family for $\operatorname{Re}\lambda > 0$. Next we can extend this family for $\operatorname{Re}\lambda > -2$ and so on. The union of these continuations gives the holomorphic family in the whole complex plane. This continuation fulfils the differential equation. Calculate H^0; take a function $\psi_0 \in S(\mathbb{R})$ that coincides with $\exp(-x)$ for $x > 0$. We have for $\operatorname{Re}\lambda > 0$

$$H^{\lambda}(\phi) = H^{\lambda}(\phi - \phi(0)\psi) + \phi(0)H^{\lambda}(\psi),$$

$$H^{\lambda}(\psi) = \frac{1}{\Gamma(\lambda)}\int_0^{\infty} x^{\lambda-1}\exp(-x)\,\mathrm{d}x = 1,$$

$$H^{\lambda}(\phi - \phi(0)\psi) = \frac{1}{\Gamma(\lambda)}\int x^{\lambda-1}(\phi(x) - \phi(0)\psi(x))\,\mathrm{d}x.$$

The last integral has analytic continuation for $\operatorname{Re}\lambda > -1$ since the function $\phi(x) - \phi(0)\psi(x)$ vanishes at $x = 0$. On the other hand the dominator Γ has a pole at $\lambda = 0$ and the right sides vanishes at $\lambda = 0$. Therefore $H^{\lambda}(\phi) \to \phi(0)$ as $\lambda \to 0$, i.e., $H^0 = \delta_0$. From the differential equation we find $H^{-k} = \left(H^0\right)^{(k)} = \delta_0^{(k)}$, $k = 1, 2, \ldots$. ▶

We define $H_{-}^{\lambda} = x_{-}^{\lambda-1}\mathrm{d}x/\Gamma(\lambda) \doteq (-x)_{+}^{\lambda-1}\mathrm{d}x/\Gamma(\lambda)$ for $\operatorname{Re}\lambda > 0$ and extend for all λ by $H_{-}^{\lambda}(\phi) = H^{\lambda}(\psi)$ where $\psi(x) = \phi(-x)$.

Example 5. Consider the Cauchy integral

$$\lim_{\varepsilon\to+0}\int (x \pm \varepsilon\imath)^{\lambda}\varphi(x)\mathrm{d}x$$

for a test density $\varphi \mathrm{d}x \in S(\mathbb{R})$. These limits exists for any $\lambda \in \mathbb{C}$ and defines two families of generalized functions in $\mathbb{R}$ denoted by $(x \pm 0\imath)^\lambda$, respectively. Indeed, integrating by parts k times yields

$$\int (x \pm \varepsilon\imath)^\lambda \varphi(x)\mathrm{d}x = \frac{(-1)^k}{(\lambda+1)\ldots(\lambda+k)} \int (x \pm \varepsilon\imath)^{\lambda+k} \varphi^{(k)}(x)\mathrm{d}x.$$

The right-hand side has meromorphic continuation at the half-plane $\operatorname{Re}\lambda > -k-1$ with occasional poles at $\lambda = -1, -2, \ldots, -k$ where the dominator vanishes. The numerator vanishes too which can be seen from the partial integration in the opposite direction. Therefore the right-hand side has holomorphic continuation to the half-plane and hence, to the whole plane. These generalized functions depend analytically on the parameter $\lambda \in \mathbb{C}$ since the functions $(x \pm \varepsilon\imath)^\lambda$ do. We have $\mathrm{d}(x \pm 0\imath)^\lambda/\mathrm{d}x = \lambda(x \pm 0\imath)^{\lambda-1}$. The products $(x \pm 0\imath)^\lambda \mathrm{d}x$ are tempered distributions. Two important equations hold:

$$(x+0\imath)^{-1}\mathrm{d}x + (x-0\imath)^{-1}\mathrm{d}x = 2\left[\frac{\mathrm{d}x}{x}\right],\ (x-0\imath)^{-1} - (x+0\imath)^{-1} = \mathrm{j}\delta_0. \quad (1.1)$$

Moreover, for any $\lambda \neq -1, -2, \ldots$ we have

$$(x \pm 0\imath)^\lambda = x_+^\lambda + \exp(\pm\lambda\pi\imath)\, x_-^\lambda. \quad (1.2)$$

This equation is obvious for the case $\operatorname{Re}\lambda > -1$ where both sides are locally integrable. It holds for all $\lambda \neq -k$ due to uniqueness of analytic continuation.

OPERATIONS. Linear operations: addition and multiplication by scalars are well defined in the space of distributions $D'(U)$ and in the space of generalized functions $K'(U)$ as operations on linear functionals. For any smooth function a in U the multiplication operator $u \mapsto au$ is also well defined by $au(\phi) = u(a\phi)$, where ϕ is any test function (density) in U. Partial derivatives are defined by the formula

$$\frac{\partial u}{\partial x_i}(\phi) = -u\left(\frac{\partial \phi}{\partial x_i}\right),$$

which is consistent with differentiation of smooth densities $u = f\mathrm{d}x$. Let $A = A(x, D) = \sum A_i(x) D^i$ be an arbitrary linear differential expression with smooth coefficients A_i in $U \subset \mathbf{V}$. It defines a linear operator on distributions u and on generalized functions v by

$$Au(\phi) = u(A^*\phi),\ Av(\phi\mathrm{d}x) = v(A^*(\phi)\mathrm{d}x),$$

where A^* is the (formal) *adjoint* operator. The definition of the operator A^* depends on the choice of volume density $\mathrm{d}x$ and can be read from the identity

$$\int A^*\phi\psi\mathrm{d}x = \int \phi A\psi\mathrm{d}x,$$

where ϕ, ψ are arbitrary test functions in U. An independent definition of action of a vector field T to a generalized function v can be given as

$$T(v)(\rho) = v(T^*(\rho)), T^*(\rho) = -L(T)\rho,$$

where $L(T)\rho$ is action of the Lie derivative along T on the density ρ, see Section 1.6.

Let A be again a linear differential operator in an open set U and s be a point in U. A *fundamental solution* for A with source at a point s is a generalized function F_s (or the distribution $F_s \mathrm{d}x$) in U that satisfies the equation $AF_s = \delta_s$ where δ_s is the Dirac function at s.

1.3 Tempered distributions

Consider a vector space $\mathbf{V}$ of dimension n. The Schwartz space $S(\mathbf{V})$ of test functions consists of smooth functions φ in $\mathbf{V}$ that satisfy the inequality

$$|x^p D^q \varphi(x)| \leq C(p,q), \quad x^p \doteq x_1^{p_1} \dots x_n^{p_n} \tag{1.3}$$

for any vectors $p, q \in \mathbb{Z}^n$ and some constant $C(p,q)$ in a linear coordinate system $x = (x_1, \dots, x_n)$. The convergence in $S(\mathbf{V})$ is defined by means of the system of norms

$$\|\varphi\|_k = \max_{|p|+|q| \leq k} \sup_{x \in \mathbf{V}} |x^p D^q \varphi(x)|, \; k = 0, 1, 2, \dots,$$

so that $\varphi_i \to \varphi$ if and only if $\|\varphi_i - \varphi\|_k \to 0$ as $i \to \infty$ for any $k = 0, 1, \dots$. The convergence does not depend on the choice of a coordinate system. The space $S(\mathbf{V})$ is the basis for L. Schwartz's theory of the Fourier transform of distributions. The dual space $S'(\mathbf{V})$ is the space of all linear continuous functions on $S(\mathbf{V})$; an element of this space is called a *tempered* distribution on $\mathbf{V}$.

A Schwartz density in $\mathbf{V}$ is by definition a product $\rho = \phi \mathrm{d}x$ where ϕ is a Schwartz function, $x = (x_1, \dots, x_n)$ is a linear coordinate system and $\mathrm{d}x = \mathrm{d}x_1 \wedge \dots \wedge \mathrm{d}x_n$ is a volume form in $\mathbf{V}$ [2]. We use the notation $S(\mathbf{V})\mathrm{d}x$ for the space of Schwartz densities in $\mathbf{V}$. An arbitrary element of the dual space $(S(\mathbf{V})\mathrm{d}x)'$ is called a *tempered* generalized function on $\mathbf{V}$. For an arbitrary tempered generalized function v the product $u = v\mathrm{d}x$ is a tempered distribution and any tempered distribution can be written in this form.

Examples 6. If a function f in $\mathbf{V}$ has at most polynomial growth at infinity, i.e., $|f(x)| \leq C(|x|+1)^q$ for some constants q, C, then the generalized function $[f]$ is tempered. The distributions considered in Examples 2–5 are tempered.

Fourier transform. We fix a coordinate system $x = (x_1, \dots, x_n)$ in $\mathbf{V}$; it generates the bijection $\mathbf{V} \cong \mathbb{R}^n$ and the measure (volume density) $\mathrm{d}x \doteq \mathrm{d}x_1 \wedge$

[2] It is easy to give a definition of Schwartz densities which does not depend on the choice of the specific volume form $\mathrm{d}x$.

$\cdots \wedge \mathrm{d}x_n$. For any $p \geq 1$ the space $L_p(\mathbf{V})$ is constructed by means of the measure $\mathrm{d}x$ in $\mathbf{V}$. The Fourier transform of an integrable function f is given by the integral

$$\hat{f}(\xi_1, \ldots, \xi_n) = \int_{\mathbf{V}} f(x_1, \ldots, x_n) \exp(-\mathrm{j}(\xi_1 x_1 + \cdots + \xi_n x_n))\mathrm{d}x.$$

We interpret the vector $\xi = (\xi_1, \ldots, \xi_n)$ as a point in the dual space $\mathbf{V}^*$ in such a way that the bilinear form $\langle \xi, x \rangle \doteq \xi_1 x_1 + \cdots + \xi_n x_n$ is the value of a function ξ on the vector x. We call the coordinate system $\xi_1, \ldots, \xi_n$ *dual* to the coordinate system $x_1, \ldots, x_n$ and write the Fourier integral in the abbreviated form

$$F(f\mathrm{d}x)(\xi) = \hat{f}(\xi) \doteq \int f(x) \exp(-\mathrm{j}\langle \xi, x \rangle)\mathrm{d}x.$$

The improper Riemann integral exists for each point $\xi \in \mathbf{V}^*$ and the function $\hat{f}$ is continuous and bounded:

$$|\hat{f}(\xi)| \leq \|f\|_1. \tag{1.4}$$

We write the Fourier integral as an operator applied to the density $f\mathrm{d}x$, its image is a function. This makes the operator F independent of the choice of coordinates in $\mathbf{V}$. The adjoint Fourier integral is applied to a function and is equal to a density. Therefore we write it in the form

$$F^*(g)(x) \doteq \left(\int g(\xi) \exp(\mathrm{j}\langle \xi, x \rangle)\mathrm{d}\xi \right) \mathrm{d}x,$$

where $\mathrm{d}\xi = \mathrm{d}\xi_1 \wedge \cdots \wedge \mathrm{d}\xi_n$ for the dual coordinate system $\xi_1, \ldots, \xi_n$. The operators F and F^* are inverse one to another; it will be stated in Section 1.7. Note that the operator F^* also does not depend on the choice of linear coordinates x in $\mathbf{V}$. Indeed, for another choice of coordinates y we have $y = Ax$ where A is a non-singular matrix. The dual system coordinates η in $\mathbf{V}^*$ relate to ξ by the formula $\eta y = \xi x$ which implies $\eta = A'^{-1}\xi$, where A' is a transposed matrix. This yields $\mathrm{d}y = |\det A|\,\mathrm{d}x$ and $\mathrm{d}\eta = |\det A|^{-1}\,\mathrm{d}\xi$. It follows that $\mathrm{d}\eta\,\mathrm{d}y = \mathrm{d}\xi\,\mathrm{d}x$ and

$$\left(\int \exp(\mathrm{j}\langle \eta, y \rangle)(\eta)\,\psi(\eta)\,\mathrm{d}\eta \right) \mathrm{d}y = \left(\int \exp(\mathrm{j}\langle \xi, x \rangle)(\xi)\,\psi(\xi)\,\mathrm{d}\xi \right) \mathrm{d}x,$$

q.e.d. Nevertheless, to simplify our notation we shall frequently omit the factor $\mathrm{d}x$.

Basic properties. **1.** If f is an integrable function and possesses integrable derivative $\mathrm{d}f/\mathrm{d}x_i$ for some i, $1 \leq i \leq n$, then

$$F\left(\frac{\mathrm{d}f}{\mathrm{d}x_i}\right) = \mathrm{j}\xi_i \hat{f}(\xi).$$

2. If f and $x_i f$ are integrable functions, then the function $F(f)$ has a continuous derivative in ξ_i and

$$\frac{\mathrm{d}F(f)}{\mathrm{d}\xi_i} = -\mathrm{j}F(x_i f).$$

3. The Fourier transform acts in the Schwartz spaces. Recall that $S(\mathbf{V})\,\mathrm{d}x$ means the space of Schwartz test densities in $\mathbf{V}$.

Proposition 1.2 *The Fourier transform acts continuously in the Schwartz spaces:*

$$F : S(\mathbf{V})\mathrm{d}x \to S(\mathbf{V}^*).$$

◀ It is well defined, since $S(\mathbf{V}) \subset L_1$. Now we state the inclusion $F(S(\mathbf{V})\,\mathrm{d}x) \subset S(\mathbf{V}^*)$. Take an arbitrary function $\varphi \in S$. Any derivative $\varphi^{(q)}$ belongs to S, hence by the property **1** the function $\hat{\varphi}$ is equal to $O(|\xi|^{-q})$ for arbitrary q. On the other hand the function $x^p\varphi(x)$ is again in S for any p. By **2**, the function $\hat{\varphi}$ has derivatives of arbitrary order and any derivative is bounded in $\mathbf{V}^*$. Moreover, for arbitrary p, q we have

$$(-1)^{|q|}\mathrm{j}^{|q|-|p|}F\left(D_x^p x^q \varphi\right) = \xi^p D_\xi^q F(\varphi).$$

The left side is the Fourier image of a function which is a linear combination of functions $x^k\varphi^{(l)}$, $k_i \le q_i, l_i \le p_i$. These functions belong to L_1, hence the left side is bounded by (1.4). The equation implies that the function $F(\varphi)$ belongs to $S(\mathbf{V}')$. It is easy to check by means of this equation that the operator $F : S(\mathbf{V})\mathrm{d}x \to S(\mathbf{V}^*)$ is continuous. The same arguments are valid for the operator F^*. ▶

Proposition 1.2 implies that the adjoint Fourier transform makes a test density of a test function and defines a continuous operator $F^* : S(\mathbf{V}^*) \to S(\mathbf{V})\mathrm{d}x$.

Definition. The Fourier transform of tempered distributions is the dual operator

$$F' : S'(\mathbf{V}) \to (S(\mathbf{V}^*)\mathrm{d}\xi)'$$

defined by $F'(v)(\varphi) \doteq v(F(\varphi))$. The function $F'(v)$ is a tempered generalized function. The adjoint Fourier transform is defined in a similar way $F^{*\prime} : (S(\mathbf{V})\mathrm{d}x)' \to S'(\mathbf{V}^*)$; it makes a tempered distribution of a tempered generalized function. We call F' and $F^{*\prime}$ Fourier-Schwartz transforms. They are inverse one to another.

Proposition 1.3 *The Fourier-Schwartz transforms match the Fourier and adjoint Fourier transforms of test functions and make the commutative diagram*

$$\begin{array}{llcllll} F': & S'(\mathbf{V}) & \to & (S(\mathbf{V}^*)\mathrm{d}\xi)' & F^{*\prime}: & (S(\mathbf{V}^*)\mathrm{d}\xi)' & \to & S'(\mathbf{V}) \\ & \cup & & \cup & & \cup & & \cup \\ F: & S(\mathbf{V})\,\mathrm{d}x & \to & S(\mathbf{V}^*) & F^*: & S(\mathbf{V}^*) & \to & S(\mathbf{V})\,\mathrm{d}x. \end{array}$$

◀ Indeed, any test density $g = f\,\mathrm{d}x$ defines a tempered distribution by means of integration $[g](\varphi) = \int f\varphi\mathrm{d}x$. Then $F'(g)$ is the tempered generalized function in $\mathbf{V}^*$ defined as

$$F'(g)(\psi\mathrm{d}\xi) = g(F(\psi\mathrm{d}\xi)) = \int f(x)\,\mathrm{d}x \int \exp(-\mathrm{j}\langle\xi, x\rangle)\,\psi(\xi)\,\mathrm{d}\xi$$

$$= \int f(x)\exp(-\mathrm{j}\langle\xi, x\rangle)\,\mathrm{d}x \int \psi(\xi)\,\mathrm{d}\xi = \int \hat{f}(\xi)\,\psi(\xi)\,\mathrm{d}\xi = \hat{f}(\psi\mathrm{d}\xi).$$

We changed the order of integration by means of Fubinis theorem. A similar equation holds for $F^{*\prime}$. ▶

Example 7. Let $q : \mathbf{V} \to \mathbb{R}$ be a positive quadratic form: $q(x) = 1/2 \sum q^{ij} x_i x_j$. Then

$$F(\exp(-2\pi q(x))\mathrm{d}x) = |\det q| \exp(-2\pi q^*(\xi))$$

where $\det q = \det\{q_{ij}\}$ and $q^* \doteq 1/2 \sum q_{ij} \xi^i \xi^j$ is the dual form, i.e., the matrix q_j is the inverse to q^{ij}.

PROPERTIES:

I. The operators $F', (F^*)'$ are linear and continuous with respect to the weak convergence in $S'(\mathbf{V})$ and $S'(\mathbf{V}')$.

II. They are compatible with the Fourier and Fourier-Plancherel transforms. This can be seen from the equation $F'[f] = [F(f)]$ for an arbitrary function $f \in L_1$. To prove this equation we write

$$F'([f])(\rho) = [f](F(\rho)) = \int f(x)\mathrm{d}x \int \exp(-\mathrm{j}\langle x, \xi\rangle)\varphi \mathrm{d}\xi,$$

where we set $\rho \in S$. The density on the right side is defined in the product $\mathbb{R} \times \mathbb{R}^*$ and integrable. Therefore we can change the order of integration:

$$\begin{aligned} F'[f](\rho) &= \int f(x)\mathrm{d}x \int \exp(-\mathrm{j}\langle x, \xi\rangle)\rho(\xi)\mathrm{d}\xi \\ &= \int \left(\int f(x) \exp(-\mathrm{j}\langle \xi, x\rangle)\mathrm{d}x \right) \rho(\xi)\mathrm{d}\xi = [F(f)](\rho). \end{aligned}$$

III. The operators F', $(F^*)'$ are inverse one to another. This follows from Proposition 1.2.

The properties of the Fourier-Plancherel transform given in Section 1.4 are fulfilled also by the Fourier-Schwartz operators F' and $(F^*)'$. We shall write F, F^*instead of F' and $F^{*\prime}$

Example 8. $F'(\delta_0) = [1]$.

Example 9. The function

$$\mathbf{H}\phi(x) \doteq \frac{1}{\pi} \int \frac{\phi(x-y)\mathrm{d}y}{y} = \frac{1}{\pi}[\frac{\mathrm{d}y}{y}]\,(\phi\,(x-y))$$

is well defined and smooth for any test function ϕ and $\mathbf{H}\phi\,(x) = O\left(|x|^{-1}\right)$ at infinity. Therefore $\mathbf{H}\phi$ belongs to $L_2\,(\mathbb{R})$ and the Fourier-Plancherel transform can be applied to this function. We can write $\mathbf{H}\phi = [\mathrm{d}x/\pi x] * \phi$ where the first factor is a distribution. It is easy to check the formula

$$F\left(\mathbf{H}(\phi)(\lambda)\mathrm{d}x\right) = -\imath \operatorname{sgn}(\lambda)\hat{\phi}(\lambda), \tag{1.5}$$

which means that $F'([\mathrm{d}x/x]) = -\pi\imath\,\mathrm{sgn}\,\xi$, where $\mathrm{sgn}\,\xi = \pm 1$ for $\pm\xi > 0$. By Parseval's theorem (Section 1.9) this implies that the $\mathbf{H}$, called a Hilbert operator, is an isometry in $L_2(\mathbb{R})$. From (1.5) it follows that $\mathbf{H}^2 = -\mathbf{I}$ where $\mathbf{I}$ is the identity operator in L_2.

Example 10. We have

$$F'((x+0\imath)^{-1}\mathrm{d}x) = -\mathrm{j}\xi_+^0, \quad F'((x-0\imath)^{-1}\mathrm{d}x) = \mathrm{j}\xi_-^0$$

where $\xi_\pm^0 = 1$ for $\pm\xi > 0$ and $\xi_\pm^0 = 0$ otherwise. It follows from the previous calculations and (1.1).

Example 11. We have for any $\lambda \in \mathbb{C}$

$$F\left(H^\lambda\right) = (\mathrm{j}\xi + 0)^{-\lambda}, \; F\left(H_-^\lambda\right) = \overline{F\left(H^\lambda\right)} = (-\mathrm{j}\xi + 0)^{-\lambda}. \tag{1.6}$$

This implies the group property of Euler kernels with respect to convolution

$$H^\lambda * H^\mu = H^{\lambda+\mu}. \tag{1.7}$$

Example 12. We define the distributions $x_\pm^{-k}$ for $k = 1, 2, \ldots$ as follows

$$x_\pm^{-k} = \lim_{\lambda\to -k}\left(x_\pm^\lambda - (\pm 1)^{k-1}\Gamma(\lambda+1)\,\delta_0^{(k-1)}\right).$$

We have

$$\begin{aligned}
F\left(x_+^{-k}\mathrm{d}x\right) &= \lim_{\lambda\to -k}(\lambda+k)\,\Gamma(\lambda+1)\,\frac{1}{(\lambda+k)}\left[(\mathrm{j}(\xi - 0\imath))^{-\lambda-1} - (\mathrm{j}\xi)^{k-1}\right] \\
&= -\lim_{\lambda\to -k}(\lambda+k)\,\Gamma(\lambda+1)\,\frac{\partial}{\partial\mu}(\mathrm{j}\xi+0)^\mu\,|_{\mu=k-1} \\
&= \frac{(-1)^k}{(k-1)!}(\mathrm{j}\xi+0)^{k-1}\left[\ln(2\pi(\xi - 0\imath)) + \imath\pi/2\right]
\end{aligned}$$

and $F\left(x_-^{-k}\mathrm{d}x\right) = \overline{F\left(x_+^{-k}\mathrm{d}x\right)}$.

Example 13. The Poisson formula implies $F\left(\sum\delta_k\mathrm{d}x\right) = \sum\delta_k$. By (1.19) for an arbitrary $h \neq 0$,

$$F\left(\sum\delta_{kh}\mathrm{d}x\right) = |h|^{-1}\sum\delta_{k/h}.$$

1.4 Homogeneous distributions

Let $\mathbf{V}$ be a vector space of dimension n. A distribution or a generalized function u in $\mathbf{V}$ is said to be *homogeneous* of degree $\lambda \in \mathbb{C}$, if it satisfies the equation $L(e)\,u = \lambda u$ where $e = \sum_0^n x_i\partial/\partial x_i$ is the Euler field and $L(e)$ is the corresponding Lie derivative (see the next section). This equation means that

$$u(L(e)\,\phi) = -\lambda u(\phi)$$

for any test function ϕ and any test density ρ, respectively. This definition agrees with the classical one, since for a smooth function or distribution u and test density, respectively, function the following equation holds:

$$\int L(e)\, u\phi + \int uL(e)\, \phi = \int L(e)\, (u\phi) = 0.$$

Example 14. The density $\mathrm{d}x$ is a homogeneous distribution of degree n and the Dirac function $\delta_0(\phi\, \mathrm{d}x) = \phi(0)$ is a homogeneous generalized function of degree $-n$, the Dirac distribution $\delta_0 \mathrm{d}x$ is homogeneous of degree 0.

Any homogeneous distribution and any homogeneous generalized function is tempered.

Fourier transform

Proposition 1.4 *Let u be a homogeneous distribution in $\mathbf{V}$ of degree $\alpha \neq 0, -1, -2, \ldots$. The Fourier transform $\widehat{u}$ is a homogeneous generalized function on $\mathbf{V}^*$ of degree $\beta = -\alpha$ and can be found by means of the integral*

$$\widehat{u}(\xi) = \int_\Gamma \tau_\alpha(\langle \xi, x\rangle) e \dashv u, \tag{1.8}$$

where

$$\tau_\alpha(t) = \Gamma(\alpha) \mathrm{j}^{-\alpha} (t - 0\imath)^{-\alpha}$$

and Γ is an arbitrary cycle in $\mathbf{V} \setminus \{0\}$ that is homologically equivalent to a sphere $\mathbf{S}$ in $\mathbf{V}$ centred at the origin.

◀ Introduce a Euclidean structure in $\mathbf{V}$ and take for $\mathbf{S}$ the unit sphere. We have

$$u(\phi) = (e \dashv u)\, |\mathbf{S}\, (\psi)\,,\ \psi(\omega) = \int_0^\infty \phi(r\omega)\, r^{\alpha-1} \mathrm{d}r,$$

since u is homogeneous, where $|\mathbf{S}$ means restriction to the unit sphere. It follows that $u_\varepsilon \to u$ as $\varepsilon \to 0$ in the sense of distributions, where $u_\varepsilon(x) = \exp(-\varepsilon |x|) u(x)$. The distribution u_ε decreases fast at infinity and its Fourier transform equals

$$\widehat{u}_\varepsilon(\xi) = (e \dashv u)\, |\mathbf{s}\, (\psi_0)\,,\ \psi_0(\omega, \xi) \doteq \int_0^\infty \exp(-\mathrm{j}r \langle \xi, \omega\rangle - \varepsilon r)\, r^{\alpha-1} \mathrm{d}r.$$

Set $\theta = \langle \xi, \omega \rangle$ and use the Euler integral

$$\int_0^\infty \exp(-\mathrm{j}\theta r - \varepsilon r)\, r^{\alpha-1} \mathrm{d}r = \Gamma(\alpha)\, \mathrm{j}^{-\alpha} (\theta + \mathrm{j}\varepsilon)^{-\alpha}.$$

This yields

$$\widehat{u}_\varepsilon(\xi) = \Gamma(\alpha)\, \mathrm{j}^{-\alpha} (e \dashv u) \left((\langle \xi, \omega\rangle + \mathrm{j}\varepsilon)^{-\alpha} \right).$$

Take the limit in both sides as $\varepsilon \to 0$ and get (1.8) for the cycle $\Gamma = \mathbf{S}$. The current $\tau_\alpha e \dashv u$ is closed, since we have

$$\mathrm{d}(\tau_\alpha e \dashv u) = \mathrm{d}(e \dashv \tau_\alpha u) = L_e(\tau_\alpha u).$$

The form $\tau_\alpha u$ is homogeneous of degree 0 hence, the right side vanishes. Therefore the integral (1.8) does not depend on the choice of Γ in the cohomology class of the sphere in $\mathbf{V}\setminus\{0\}$. ▶

For an arbitrary homogeneous generalized function f in $\mathbf{V}$ of degree $\alpha > -n$, the product $u = f\,\mathrm{d}x$ is well defined as a tempered distribution. It is homogeneous of degree $\alpha + n > 0$. The Fourier transform $\widehat{f} = F(f\mathrm{d}x)$ is a homogeneous function of degree $\beta = -\alpha - n$ and the equation (1.8) reads as follows

$$\widehat{f}(\xi) = \int_\Gamma \tau_\alpha(\xi x) f(x) \omega_{\mathbf{E}}, \tag{1.9}$$

where $\omega_{\mathbf{E}} = e \dashv \mathrm{d}x$ is the projective volume form.

Important examples of homogeneous distributions are

RIESZ KERNELS. Let Q be a positive quadratic form in a space $\mathbf{V}$, say $q(x) = 1/2\sum q_{ij}x_i x_j$. The quadratic form $q^*(\xi) = 1/2\sum q^{ij}\xi_i\xi_j$ (called dual quadratic form) is defined in the dual space $\mathbf{V}^*$, where the matrix $\{q^{ij}\}$ is inverse to $\{q_{ij}\}$. It is also positive. Consider the family of densities (Riesz kernels)

$$R(\lambda) \doteq \frac{(2\pi q)^{\lambda - n/2}\,\mathrm{d}x}{\Gamma(\lambda)}$$

in $\mathbf{V}$, where $\mathrm{d}x = \mathrm{d}x_1 \wedge \cdots \wedge \mathrm{d}x_n$ for a coordinate system $x = (x_1, \ldots, x_n)$ in $\mathbf{V}$. This density is locally integrable if $\operatorname{Re}\lambda > 0$.

Proposition 1.5 *The family $\lambda \mapsto R(\lambda)$ admits analytic continuation on the complex plane The distribution $R(\lambda)$ is homogeneous of degree 2λ.*

◀ For a proof we can apply the method of Proposition 1.1 and the identity

$$q^*(D)\, q^\lambda = \lambda(\lambda - 1 + n/2)\, q^{\lambda-1}, \tag{1.10}$$

where $q^*(D) = 1/2\sum q^{ij}\partial^2/\partial x_i \partial x_j$. Check the second statement: for $\operatorname{Re}\lambda > n/2 + 1$ the density $q^{\lambda - n/2}\mathrm{d}x$ is homogeneous and the equation $L(e)(R(\lambda)) = 2\lambda R(\lambda)$ holds in the classical sense. This equation holds for all λ due to uniqueness of analytic continuation. ▶

The function

$$Q(\lambda) \doteq \frac{(2\pi q)^\lambda(x)}{\Gamma(\lambda + n/2)}$$

is locally integrable, if $\operatorname{Re}\lambda > -n/2$ and defines a generalized function in $\mathbf{V}$. The family $\{Q(\lambda)\}$ also admits a holomorphic continuation at $\mathbb{C}$. For any $\lambda \in \mathbb{C}$,

$Q(\lambda)$ is a homogeneous tempered generalized function of degree 2λ, (called a Riesz function).

Exercise. Check the formula

$$R(-k) = \frac{1}{\Gamma(k+n/2)}\left(-\frac{1}{2\pi}q^*(D)\right)^k \delta_0(x)\,\mathrm{d}x,\ k=0,1,2,\dots$$

Define the family distribution $R^*(\lambda) = (2\pi q^*)^{\lambda-n/2}\,\mathrm{d}x/\Gamma(\lambda)$ and of generalized functions $Q(\lambda) = (2\pi q^*)^\lambda/\Gamma(\lambda+n/2)$ in the dual space $\mathbf{V}^*$ by means of the dual quadratic form q^*. Take analytic continuation of these families at $\mathbb{C}$.

Proposition 1.6 *For any $\lambda \in \mathbb{C}$ the Fourier-Schwartz transform of the Riesz kernel of degree λ is the dual Riesz function of degree $-\lambda$ and vice versa:*

$$F(R(\lambda)) = Q^*(-\lambda),\ F(Q(\lambda)) = R^*(-\lambda). \tag{1.11}$$

In particular,

$$F\left(\frac{(2\pi q)^{\lambda-n/2}\,\mathrm{d}x}{\Gamma(\lambda)}\right) = \frac{(2\pi q^*(\xi))^{-\lambda}}{\Gamma(n/2-\lambda)}$$

for all λ such that $n/2 > \operatorname{Re}\lambda > 0$.

◀ Introduce the Euclidean structure in $\mathbf{V}$ such that $q(x) = |x|^2$; then we have $q^*(\xi) = |\xi|^2$. Apply Proposition 1.4 to the homogeneous distribution u of degree $\alpha = 2\lambda$ such that $u|\mathbf{S} = \mathbf{1}$. Then we only need to evaluate the integral

$$I_\lambda(\xi) = \int_{\mathbf{S}} (\langle \xi, \omega\rangle + 0\imath)^{2\lambda}\,\mathrm{d}\omega,$$

where $\mathrm{d}\omega = e \dashv \mathrm{d}x$ is the Euclidean area on the unit sphere $\mathbf{S}$. The integral is spherically symmetric and reduces to a Beta function:

$$I(\xi) = s_{n-1}(1+\exp(\mathrm{j}\lambda))\,2^{-1}B(1/2-\lambda,(n-1)/2)\,|\xi|^{-2\lambda},$$

where s_{n-1} is the area of the unit sphere in $\mathbf{E}^{n-1}$. This implies the first equation (1.11). The second equation follows from the first one. ▶

Powers of arbitrary quadratic form. Let q be an arbitrary real non-singular quadratic form in $\mathbf{V}$. One can define the family of locally integrable functions $Q_+^\lambda = q_+^\lambda/\Gamma(\lambda+1)\Gamma(\lambda+n/2)$ where $\lambda \in \mathbb{C}$, $\operatorname{Re}\lambda > -1$. This family can be extended to the whole plane $\mathbb{C}$ by means of the identity (1.10) which holds for any quadratic form q. This implies that the family of functions q_+^λ has meromorphic continuation with occasional poles at the points $\lambda = -1,-2,\dots$ and $\lambda = -n/2, -n/2-1, \dots$. Another method is similar to that of Example 5. We define powers of q as follows:

$$(q \pm 0\imath)^\lambda(\rho) \doteq \lim_{\varepsilon\to+0}\int (q\pm\varepsilon\imath)^\lambda \rho,\ \rho \in K(\mathbf{V}). \tag{1.12}$$

The power $(q \pm \varepsilon \imath)^{\lambda}$ is defined by means of the branch of the power function $\zeta^{\lambda} \doteq \exp(\lambda \ln \zeta)$, where $\ln \zeta$ is the branch of the logarithm in $\mathbb{C} \backslash (-\mathbb{R}_{+})$ that is real for positive ζ.

Proposition 1.7 *The limits* (1.12) *exist for any* ρ *and define holomorphic families* $\lambda \mapsto (q \pm 0\imath)^{\lambda}$ $\lambda \in \mathbb{C}$, *of generalized functions in* $\mathbf{V}$.

◀ Choose a positive quadratic form $p = p(x)$ in $\mathbf{V}$ Fix $\varepsilon > 0$ and consider the family of functions $q_{\varepsilon}^{\lambda}$ where $q_{\varepsilon}(x) = q(x) + \varepsilon p$. It is holomorphic for $\lambda \in \mathbb{C}$ since q_{ε} does not vanish in $\mathbf{V}$. It fulfils the identity $q_{\varepsilon}^{*}(D) q_{\varepsilon}^{\lambda}(x) = \lambda(\lambda - 1 + n/2) q_{\varepsilon}^{\lambda - 1}(x)$ where q_{ε}^{*} is the homogeneous second-order operator dual to the form q_{ε}. Therefore we have identity

$$q_{\varepsilon}^{\lambda-1}(\rho) = \frac{1}{\lambda(\lambda - 1 + n/2)} q_{\varepsilon}^{\lambda}(q_{\varepsilon}^{*}(D)\rho). \tag{1.13}$$

The right-hand side has meromorphic continuation to the half-plane $\{\lambda; \mathrm{Re}\lambda > -1\}$ with occasional pole $\lambda = 0$, since the dominator vanishes. Integrating by parts yields that the numerator also vanishes as $\lambda = 0$. Therefore there is no pole at this point and we get analytic extension of both sides to the half-plane $\{\mathrm{Re}\,\lambda > -1\}$. Using again (1.13) gives analytic continuation at $\{\lambda; \mathrm{Re}\,\lambda > -2\}$ and so on. Another occasional pole is $\lambda = 1 - n/2$. Integrating by parts yields again

$$\int q_{\varepsilon}^{1-n/2}(q_{\varepsilon}^{*}(D)\rho) = 0,$$

which means that there is no pole at all. Now we take $\varepsilon \to 0$; the dual form q_{ε}^{*} tends to the dual form q^{*} since q is non-singular. The right-hand side of (1.13) has a limit for $\mathrm{Re}\,\lambda > -1$ hence, the left side also does. Repeating the above arguments, shows that the family $q_{\varepsilon}^{\lambda}$ has a limit as $\varepsilon \to 0$. It coincides with the function $(q + 0\imath)^{\lambda}$ for $\mathrm{Re}\,\lambda > 0$. This implies that the family $(q + 0\imath)^{\lambda}$ has analytic continuation on the whole plane. The same is true for $(q - 0\imath)^{\lambda}$. ▶

Comparing the power functions, we get $(q \pm 0\imath)^{\lambda} = q_{+}^{\lambda} + \exp(\pm\pi\lambda\imath)\, q_{-}^{\lambda}$, where $q_{-}^{\lambda} = (-q)_{+}^{\lambda}$.

For an arbitrary natural integer k we define the generalized function

$$\left[q^{-k}\right] = \frac{1}{2}\left[(q + 0\imath)^{-k} + (q - 0\imath)^{-k}\right].$$

Note the equation

$$\delta^{(k-1)}(q) = \frac{(n-1)!}{\mathrm{j}}\left[(q - 0\imath)^{-k} - (q - 0\imath)^{-k}\right]. \tag{1.14}$$

1.5 Manifolds and differential forms

DIFFERENTIAL FORMS AND ORIENTATION. Let $\mathbf{V}$ be a vector space and $x = (x_1, \ldots, x_n)$ be a coordinate system in $\mathbf{V}$. The differentials $\mathrm{d}x_1, \ldots, \mathrm{d}x_n$ are differential forms of degree 1 in $\mathbf{V}$. For an open set $U \subset \mathbf{V}$ and consider the algebra $O(U)$ of infinitely differentiable functions in U. Take the exterior algebra $\Omega(U)$ generated by the forms $\mathrm{d}x_1, \ldots, \mathrm{d}x_n$ over $O(U)$; it is called the algebra of *even differential forms* in U. Any even form α of degree k is an element of $\Omega(U)$; it can be written as

$$\alpha = \sum a_{i_1,\ldots,i_k}(x)\, \mathrm{d}x_{i_1} \wedge \cdots \wedge \mathrm{d}x_{i_k}, \tag{1.15}$$

where $\wedge$ is the symbol of the exterior product. This operation is bilinear and fulfils the condition $\beta \wedge \alpha = \pm\alpha \wedge \beta$ where the sign is $-$ if both forms α, β have odd degrees and $+$ otherwise. The coefficients $a_{i_1,\ldots,i_k}$ belong to $O(U)$; they are uniquely defined if the sum (1.15) is taken only over the set of indices such that $i_1 < \cdots < i_k$. To write an even form in another coordinate system $y = (y_1, \ldots, y_n)$, we need only to replace each generator of the algebra $\Omega(U)$ by means of the standard rule

$$\mathrm{d}x_i = \sum \frac{\partial x_i}{\partial y_j} \mathrm{d}y_j, \; i = 1, \ldots, n. \tag{1.16}$$

Definition. Let X be a smooth manifold; the structure of the manifold is given by a set (atlas) of maps (U, x), where U is an open set in X and $x = (x_1, \ldots, x_n)$ is a coordinate system in U. An even differential form in X is a form α_U defined for each map (U, x) in X such that for any two maps $(U, x), (V, y)$ the forms α_U and α_V are related in $U \cap V$ by the rule (1.16). Let $\Omega(X)$ be the space of even differential forms in X. The exterior differential of the form α of degree k is the globally defined form $\mathrm{d}\alpha$ of degree $k+1$, defined locally as

$$\mathrm{d}\alpha = \sum \mathrm{d}a_{i_1,\ldots,i_k} \wedge \mathrm{d}x_{i_1} \wedge \cdots \wedge \mathrm{d}x_{i_k}.$$

Definition. An *orientation* on a manifold X is a choice of an atlas of maps such that the Jacobian $\det \partial y/\partial x$ is always positive for any two maps (U, x), (V, y) in the atlas. If X is an oriented manifold of dimension n, an *orientation* form on X is an arbitrary even n-form σ on X such that $\sigma(\partial/\partial x_1, \ldots, \partial/\partial x_n) > 0$ for any coordinate system $x_1, \ldots, x_n$ that belongs to the atlas of the orientation.

Example 15. Let $\mathbf{V}$ be a vector space of dimension n. The set of one-dimensional subspaces $L \subset \mathbf{V}$ has the structure of an (algebraic) manifold of dimension $n-1$; it will be denoted $\mathbb{P}(\mathbf{V})$ (the projective space of $\mathbf{V}$). Take a system of coordinates $x_1, \ldots, x_n$ in $\mathbf{V}$; the set U_j of lines L such that $x_j \neq 0$ in L is an open set in $\mathbb{P}(\mathbf{V})$. The coordinates $x_1, \ldots, x_{j-1}, x_{j+1}, \ldots, x_n$ of the intersection point $y(L) \doteq L \cap \{x_j = 1\}$ form a local coordinate system in $U_j, j = 1, \ldots, n$. The manifold $\mathbb{P}(\mathbf{V})$ is orientable if n is even and is not orientable if n is odd.

INTEGRAL. Let X be an oriented manifold; the integral $\int_X \alpha$ is well defined for any even differential form on X with compact support and continuous coefficients. For any even form α the "Stokes" integral formula holds,

$$\int_U \mathrm{d}\alpha = \int_{\partial U} \alpha,$$

where ∂U is the smooth boundary of a compact oriented open set $U \subset X$. The orientation of ∂U is defined by means of the form $\mathrm{d}f$ where f is a function that fulfils the conditions: $f < 0$ in U and $f = 0$, $\mathrm{d}f \neq 0$ on ∂U. This means that a local coordinate system $y = (y_2, \ldots, y_n)$ belongs to the atlas of the orientation of U, if the coordinate system $y_1 = f(x), z_2, \ldots, z_n$ belongs to the atlas of orientation of U and $y_j = z_j | \partial U$, $j = 2, \ldots, n$.

Definition. The form α is called *odd*, if there is one more factor $\operatorname{sgn}\det \partial y / \partial x$ in the transformation formula from coordinates y to coordinates x. Any odd form with compact support can be integrated over the manifold X with no orientation. A *volume* form α in X is an odd form such that $\alpha = a\mathrm{d}x_1 \wedge \cdots \wedge \mathrm{d}x_n$ and $a \neq 0$ for any local coordinate system. There is no Stokes theorem for odd forms. See [18] for more information.

Example 16. A measure, a charge, a density, a surface density = area element, a curve element are all odd forms.

CRITICAL POINTS. Let $f : X \to Y$ be a smooth mapping of manifolds of local dimensions $\dim(X, x), \dim(Y, y)$. A point $x_0 \in X$ is called *critical*, if the rank of the linear mapping $\mathrm{d}f(x_0)$ is less than $\dim(Y, f(x_0))$ Note that the mapping $\mathrm{d}f$ (x_0) is given by the Jacobian matrix $\{\partial y_j / \partial x_i\}$ in local coordinate system $\{x_i\}$ at x_0 and $\{y_j\}$ at $f(x_0)$ respectively, and $\operatorname{rank} \mathrm{d}f\ (x_0) = \operatorname{rank}\{\partial y_j / \partial x_i\}$. A point $y_0 \in Y$ is called a *critical value* of f if $y_0 = f(x_0)$ for, at least one critical point x_0. By Sard's theorem the set of critical values has zero measure in Y.

DEGREE OF A MAPPING. A mapping $f : X \to Y$ of manifolds (or topological spaces) is called *proper*, if for any compact set K in Y the set $f^{-1}(K)$ is compact in X. Let f be a proper mapping of smooth oriented manifolds of the same dimension $n = \dim X = \dim Y$. Take an arbitrary non-critical point $z \in Y$ and consider the sum

$$\deg(f, z) \doteq \sum \operatorname{sgn} \frac{\partial y}{\partial x} |_{f(x)=z}. \tag{1.17}$$

The sum is finite since the mapping f is proper and each sign is well defined since z has non-critical value.

Proposition 1.8 *If Y is connected, the number* $\deg(f, z)$ *is constant.*

◀ If we move the point z along a generic curve in Y, then either each term in (1.17) stays constant or two terms 1 and -1 disappear or appear simultaneously. ▶

The number (1.17) is called the *degree* of the mapping f and will be denoted by $\deg(f)$. The degree depends on the orientations and changes its sign if an orientation is changed to the opposite one. The basic property of the degree is as follows: if α is an even differential n-form on Y with compact support (more generally, an integrable form on Y), then

$$\int_X f^*(\alpha) = \deg(f) \int_Y \alpha.$$

Example 17. Consider the mapping $p : \mathbb{C} \to \mathbb{C}$ defined by a complex polynomial $p = p(z)$. Critical points of p are the roots of p'. The degree of this mapping coincides with the degree of the polynomial p.

TANGENT VECTORS AND COVECTORS. Let x be a point in a smooth manifold X, O_x be the algebra of smooth functions defined in a neighborhood of x. Any functional $t : O_x \to \mathbb{R}$ satisfying the equation $t(ab) = t(a)\,b(x) + a(x)\,t(b)\,, a, b \in O_x$ is called a *tangent* vector in X at x. The space of all tangent vectors at x is denoted T_x (tangent space at x). The dual vector space T_x^* is called the *cotangent* space at x. An element τ of the cotangent space is called a differential form of degree 1 or covector at x. Let Y be a submanifold of X. A vector t is tangent to Y if $t(a) = 0$ for any function $a \in O_x$ that vanishes in Y. Any functional $\tau \in T_x^*$ such that $\tau(t) = 0$ for any tangent vector t to Y is called a *conormal* covector to Y at x. The union of spaces $T(X) \doteq \cup T_x$, $T^*(X) \doteq \cup T_x^*$ are called the *tangent* bundle and the *cotangent* bundle, respectively. If X is an open set in a space $\mathbf{V} \doteq \mathbb{R}^n$, the tangent bundle is isomorphic to $X \times \mathbf{V}$ and the cotangent bundle has the structure $T^*(X) = X \times \mathbf{V}^*$ where $\mathbf{V}^*$ is the space dual to $\mathbf{V}$. Let $O(X)$ be the algebra of smooth functions defined in X. A linear operator t in this algebra is called a *tangent field*, if it satisfies the Leibniz condition: $t(ab) = t(a)\,b + at(b)$. Any tangent field t defines at each point $x \in X$ the tangent vector $t_x : O_x \to \mathbb{R}$ such that $t_x(a) = t(a)$ for functions a defined on X. A differential form α of degree k at a point $x \in X$ is a mapping $T_x \times \cdots \times T_x \to \mathbb{C}$, i.e., $\alpha(t_1, \ldots, t_n) \in \mathbb{C}$ that is linear in each argument and changes its sign if two neighboring arguments permute. If the form α is given by (1.15), then

$$\alpha(t_1, \ldots, t_k) = a_{i_1,\ldots,i_k} \sum (-1)^\pi \, dx_{i_1}(s_1) \wedge \cdots \wedge dx_{i_k}(s_k)\,, \tag{1.18}$$

where the inner sum is taken over all permutations $s_1, \ldots, s_k$ of the vectors $t_1, \ldots, t_k$ and π is the parity of the permutation. Vice versa, any differential form at a point x can be written as the sum (1.15) with some coefficients $a_{\ldots} \in \mathbb{C}$. A differential form α of degree k on the manifold X is an arbitrary smooth family of differential forms α_x, $x \in X$ of degree k; the condition of smoothness means that the coefficients $a_{\ldots}$ in (1.18) are smooth functions of x in any chart.

CONTRACTION. Let s be a tangent field in X and α be a differential form of degree $k > 0$. The contraction of α by means of s is the differential form β or degree $k-1$, denoted by $s \vdash \alpha$, such that

$$\beta(t_1, \ldots, t_{k-1}) = \alpha(s, t_1, \ldots, t_{k-1})\,.$$

We have always $s \vdash (s \vdash \alpha) = 0$ and $t_1 \vdash (t_2 \vdash \ldots (t_k \vdash \alpha)) = \alpha(t_1, \ldots, t_k)$.

Example 18. Let $x_1,\ldots,x_n$ be a coordinate system in a space $\mathbf{V}$ and $e = \sum x_i \partial/\partial x_i$ be the Euler tangent field. Then

$$e \dashv \mathrm{d}x_1 \wedge \cdots \wedge \mathrm{d}x_n = \sum (-1)^{i-1} x_i \mathrm{d}x_1 \wedge \cdots \wedge \mathrm{d}x_{i-1} \wedge \mathrm{d}x_{i+1} \wedge \cdots \wedge \mathrm{d}x_n.$$

Given a tangent field v, there is defined the *Lie derivative* $L(e)$ along v of an arbitrary even differential form α,

$$L(v)\alpha = \mathrm{d}(v \vdash \alpha) + v \vdash \mathrm{d}\alpha.$$

The operator $L(v)$ preserves the degree of the form and commutes with the exterior differential. For a function a (form of zero degree) the Lie derivative coincides with the action of the tangent field: $L(v)a = v(a)$.

QUOTIENT OF FORMS. A form β of degree k is called non-degenerated at x if $\beta(t_1,\ldots,t_k) \neq 0$ for some tangent vectors at x. Let β be a non-degenerated form and α be an arbitrary differential form of degree $n = \dim X$. The Leray form or *quotient* α/β is an arbitrary $n-k$-form γ such that $\beta \wedge \gamma = \alpha$. Let $f_1,\ldots,f_k$ be some smooth functions in X such that the form $\beta = \mathrm{d}f_1 \wedge \cdots \wedge \mathrm{d}f_k$ is non-degenerated. Then the quotient α/β is uniquely defined up an additive term of the form $\mathrm{d}f_1 \wedge \beta_1 + \cdots + \mathrm{d}f_k \wedge \beta_k$.

Definition. A manifold X is called a Riemannian manifold if there is given a smooth positive symmetric bilinear form g (called a metric) on the tangent bundle $T(X)$. These conditions mean that for any tangent vectors u, v at a point x we have $g(v,u) = g(u,v)$, $g(u,u) > 0$, if $u \neq 0$, and for any coordinate system $x = (x_1,\ldots,x_n)$ in an open set $U \subset X$ the coefficients $g_{ij}(x)$ in the local expression $g = \sum g_{ij}(x)\,\mathrm{d}x_i \mathrm{d}x_j$ are smooth functions in U. For any C^1-curve γ given by the parametric equation $x = x(t)$, $0 \le t \le 1$, the length $|\gamma|$ is defined to be

$$|\gamma| = \int_0^1 g^{1/2}(x_t', x_t')\,\mathrm{d}t = \sum \int_0^1 g_{ij} \frac{\partial x_i}{\partial t} \frac{\partial x_j}{\partial t} \mathrm{d}t.$$

For any points $y, z \in X$ the distance $\mathrm{dist}(y,z)$ is the length of the shortest curve that joins these points. A geodesic curve in X is a curve γ which is locally a shortest way between any two points $y, z \in \gamma$ whose distance is sufficiently small.

The Riemannian metric g gives sense to the length of a vector, to the angle between two vectors and induces the area density on any smooth submanifold $Y \subset X$. A submanifold Y is called totally geodesic, if any geodesic curve that is tangent to Y is contained in Y.

See more information in [98] or other textbooks on differential geometry.

1.6 Push down and pull back

Distributions and generalized functions have different behavior under change of variables. Take a smooth invertible mapping $f : U \to \mathbf{V}$ defined in an open set $U \subset \mathbf{W}$. Then any test function $\psi \in D(\mathbf{V})$ is transformed to a test function

$\phi \in D(U)$ by $\phi(x) = f^*(\psi)(x) \doteq \psi(f(x))$. The pull back (inverse image) of a test density $\rho = \psi \mathrm{d}x$ is equal to $\sigma = f^*(\rho) \doteq \phi |\det \partial y/\partial x| \,\mathrm{d}x$ where $\partial y/\partial x$ is the Jacobian of the mapping f. The push down (direct image) of a distribution u and pull back (inverse image) of a generalized function v are defined in the natural way: for $u \in D'(U), v \in K'(U)$ we set

$$f_*(u)(\psi) = u(f^*(\psi)), \ f^*(v)(\rho) = v(g^*(\rho)), \text{ where } g = f^{-1}. \tag{1.19}$$

If $v = v(x)$ is a locally integrable function, we have $f^*(v)(x) = v(f(x))$.

Example 19. Take a linear invertible transformation $A : \mathbb{R}^n \to \mathbb{R}^n$. For the Dirac distribution $\delta_0 \mathrm{d}x$ we have $A_*(\delta_0 \mathrm{d}x) = \delta_0 \mathrm{d}y$, unlike the equation $A^*(\delta_0) = |\det A|^{-1} \delta_0$ for the Dirac function.

These constructions are generalized for more general mappings of smooth manifolds and distributions and generalized functions defined on such manifolds. In the general case, a generalized function may have pull back, whereas a distribution may have pull down. We consider here only the first operation.

PULL DOWN OF A TEST DENSITY. Let X, Y be smooth oriented manifolds and $f : X \to Y$ be a submersion, i.e., a smooth mapping without critical points, i.e., such that $\operatorname{rank} \mathrm{d}f(x) = \dim(Y, f(x))$ for any $x \in X$. Take an orientation form σ in X, an orientation form ρ in Y and consider the pull back $f^*(\rho)$; this is a non-degenerated form in X of degree $\dim Y$. Take a test density $\varphi \in K(X)$ and consider the quotient $\varphi / f^*(\rho)$; it is a differential form of degree $\dim X - \dim Y$. Each *fibre* $X_y = f^{-1}(y)$ is a smooth manifold of dimension $\dim X - \dim Y$, since f is a submersion. Define the orientation of X_y by means of the volume form $\sigma / f^*(\rho)$ and consider the integral

$$\psi(y) = \int_{X_y} \frac{\varphi}{f^*(\rho)}.$$

The integral converges, since the support of φ is a compact set. The function ψ is smooth in Y, since f is submersion. The product $f_*(\varphi) \doteq \psi(y)\rho$ is defined in Y and does not depend on the choice of ρ (but depends on orientations): for another choice ρ' we have $\rho' = a\rho$ where $a > 0$ and $\varphi/\rho' = a^{-1}\varphi/\rho$, $\psi' = a^{-1}\psi$, $\psi'\rho' = \psi\rho$. The form $f_*(\varphi)$ is called the *push down* of φ. We can say in a non-formal way: push down of a form is the result of integration of the form along fibres of the mapping. It is easy to check that the form $f_*(\varphi)$ is smooth and has compact support, i.e., $f_*(\varphi) \in K(Y)$.

PULL BACK OF A GENERALIZED FUNCTION. For an arbitrary generalized function v in Y we set

$$f^*(v)(\varphi) \doteq v(f_*(\varphi)).$$

The functional $u = f^*(v)$ is well defined and continuous on the space $K(X)$, i.e., u is a generalized function in X. It is called the *pull back* of v.

Example 20. Let $\mathbf{V}$ be a vector space and U be an open subset of $\mathbf{V}$. Take a submersion $f : X \to \mathbb{R}$ and the Dirac function $v = \delta_0$. By the above definition the pull back is denoted $\delta(f) \doteq f^*(\delta_0)$, is well defined and

$$\delta(f)(\varphi) = \int_{f=0} \frac{\varphi}{\mathrm{d}f} = \lim_{\varepsilon \to +0} \frac{1}{2\varepsilon} \int_{|f| \le \varepsilon} \varphi.$$

Here $t = f(x)$ is the coordinate form of the mapping f, $\mathrm{d}t$ is the orientation form in $\mathbb{R}$. Formally $\mathrm{d}f = f^*(\mathrm{d}t)$. If X is an open set in a Euclidean space $\mathbf{E}$, $\mathrm{d}V$ is the volume form in $\mathbf{E}$ and $\varphi = \alpha\,\mathrm{d}V$, then we can write

$$\delta(f)(\varphi) = \int_{f=0} \frac{\alpha}{|\nabla f|} \mathrm{d}S,$$

where $\mathrm{d}S$ is the hypersurface area form in $\mathbf{E}$. If we replace f by the function $g = af$, where $a \ne 0$ is a smooth function, then

$$\delta(g) = \frac{1}{|a|} \delta(f).$$

Pull backs of derivatives of the Dirac function can be written in a similar way: taking $v = \delta_0^{(k)}$ gives the equation

$$\delta^{(k)}(f) \doteq \left(\frac{\mathrm{d}}{\mathrm{d}t}\right)^k \int_{f=t} \frac{\varphi}{\mathrm{d}f} |_{t=0}$$

which agrees with (1.14). Replacing f by $g = af$ for a constant $a \ne 0$, gives

$$\delta^{(k)}(g) = \frac{1}{|a|\, a^k} \delta^{(k)}(f).$$

Example 21. Let f be a real smooth function in an open set $U \subset \mathbf{V}$ such that $\mathrm{d}f \ne 0$ as $f = 0$. We define the generalized Euler kernel generated by f as

$$H_f^\lambda(\rho) = \frac{f_+^{\lambda-1}\mathrm{d}x}{\Gamma(\lambda)}(\rho) \doteq \frac{1}{\Gamma(\lambda)} \int_{f>0} f^{\lambda-1}\phi \mathrm{d}x,\ \phi \in D(U),\ \operatorname{Re}\lambda > 0.$$

In the case $n = 1$, $f = x$ this is just the Euler kernel as in Section 1.2. It possesses similar properties:

Proposition 1.9 *The family H_f^λ has analytic continuation on $\mathbb{C}$ with values in $K'(U)$. The functional $\delta^{(k)}(f) \doteq H_f^{-k}$ is supported by the hypersurface $Z \doteq \{f = 0\}$ for $k = 0, 1, 2, \ldots$.*

◀ Define the function

$$r(t) \doteq \int_{f=t} \frac{\phi \mathrm{d}x}{\mathrm{d}f},$$

where the hypersurface is oriented by the form $\mathrm{d}f$. The function r has compact support, since ρ has such support. It belongs to C^{∞} in a neighborhood of the point $t = 0$, since $\mathrm{d}f \neq 0$. Therefore we can apply the Euler distribution to the test function r. By Fubini's theorem,

$$H^{\lambda}(r) = H_f^{\lambda}(\phi), \ \operatorname{Re}\lambda > 0.$$

The left side has analytic continuation in the complex plane. This gives analytic continuation for H_f^{λ}. We have

$$H_f^{-k}(\phi) = \delta^{(k)}(r) = r^{(k)}(0) = \frac{\mathrm{d}^k}{\mathrm{d}t^k} \int_{\phi = t} \frac{\phi \mathrm{d}x}{\mathrm{d}f} |_{t=0}.$$ ▶

1.7 More on the Fourier transform

In this section, knowledge of rudiments of Lebesgue integration theory is assumed. Let X be a space supplied with a Lebesgue measure $\mathrm{d}x$. A (measurable) function $f : X \to \mathbb{C}$ is called *Lebesgue* integrable in $(X, \mathrm{d}x)$, if the Lebesgue integral

$$\|f\|_1 \doteq \int |f| \, \mathrm{d}x$$

is finite. A function f is called negligible if $f = 0$ almost everywhere; this condition is equivalent to the equation $\|f\|_1 = 0$. Any equation like $f = g$ in Lebesgue's theory means only that $f - g$ is a negligible function.

For an arbitrary number $p \geq 1$ the notation $L_p = L_p(X), p \geq 1$ is used for the $\mathbb{C}$-vector space of functions in X such that the function $|f|^p$ is Lebesgue integrable. More accurately, this is the quotient space of such functions modulo the subspace of negligible functions. This is a Banach space with the norm

$$\|f\|_p = \left(\int |f|^p \mathrm{d}x \right)^{1/p}.$$

In particular, the space $L_2(X)$ of square-integrable functions in X, is a Hilbert space with the scalar (inner) product

$$\langle f, g \rangle = \int f \, \bar{g} \, \mathrm{d}x.$$

Remind the inequality: $|\langle f, g \rangle|^2 \leq \|f\|^2 \|g\|^2$, which holds for any scalar product. It is equivalent to the triangle inequality $\|f + g\|^2 \leq (\|f\| + \|g\|)^2$ for L_2-norms.

We need few more facts from the Lebesgue theory:

Theorem 1.10 (Dominated convergence theorem) *Let F be an integrable function in V and $f_i, i = 1, 2, \ldots$ a sequence of (measurable) functions such that $|f_i| \leq F$ and $f_i \to f$ almost everywhere in X. Then*

$$\int_V f_i \, \mathrm{d}x \to \int_V f \, \mathrm{d}x.$$

Theorem 1.11 (Fubini theorem) *Let X, Y be spaces endowed with the Lebesgue measures* $\mathrm{d}x, \mathrm{d}y$, *respectively, and f be a function in $X \times Y$ integrable with respect to the measure* $\mathrm{d}x\mathrm{d}y$ *in $X \times Y$. Then the function $f(\cdot, y)$ is integrable in X for almost all $y \in Y$, the function $g(y) \doteq \int_X f(x, y)\,\mathrm{d}x$ is integrable in Y and*

$$\int_Y \mathrm{d}y \int_X f(x, y)\,\mathrm{d}x = \int_{X\times Y} f(x, y)\,\mathrm{d}x\mathrm{d}y.$$

From Fubini's theorem we conclude that

$$\int \left(\int f(x, y)\mathrm{d}x \right) \mathrm{d}y = \int \left(\int f(x, y)\mathrm{d}y \right) \mathrm{d}x,$$

i.e., we may change the order of integration for any integrable f in $X \times Y$.

Let V be a finite-dimensional vector space. Fix a system of coordinates and consider the corresponding volume form $\mathrm{d}x$ in $\mathbf{V}$. Lebesgue theory is then applicable in $\mathbf{V}$. We say that a function g in $\mathbf{V}$ is a *step* function, if it is equal to a linear combination of indicators of cubes $Q \subset V$. Recall that the indicator function χ of a set G is equal to 1 in G and 0 otherwise.

Theorem 1.12 (Density theorem) *For any $p \geq 1$ the set of step functions is dense in the space $L_p(\mathbf{V})$.*

The Fourier transform of a function $f \in L_1(\mathbf{V})$ is defined as in Section 1.1.

Theorem 1.13 (Parseval) *For an arbitrary function $f \in L_1(\mathbf{V}) \cap L_2(\mathbf{V})$ the Fourier image $\hat{f}$ belongs to $L_2(\mathbf{V}^*)$. Then the equations*

$$\|f\|^2 = \|\hat{f}\|^2, \tag{1.20}$$

$$\langle f, g \rangle = \langle \hat{f}, \hat{g} \rangle \tag{1.21}$$

hold for arbitrary $g \in L_1 \cap L_2$.

In other words, the operator $F : f \mapsto \hat{f}$ is an isometry in $L_2(\mathbf{V}) \to L_2(\mathbf{V}^*)$.

◀ Consider the case $\mathbf{V} = \mathbb{R}$ for simplicity. Take a finite interval, say $I = [a, b] \subset \mathbb{R}$ and denote by h_I the indicator of this interval. First we check (1.21) for $f = h_I, g = h_J$ for arbitrary intervals I, J. We have

$$\begin{aligned}\hat{h}_I &= \int_a^b \exp(-\mathrm{j}\xi x)\mathrm{d}x = -\frac{1}{\mathrm{j}\xi}\left[\exp(-\mathrm{j}b\xi) - \exp(-\mathrm{j}a\xi)\right] \\ &= -\exp\left(-\mathrm{j}\xi\frac{a+b}{2}\right)\frac{\sin(\pi\xi(b-a))}{\pi\xi}.\end{aligned}$$

We see that the right side is a holomorphic function in the whole plane $\mathbb{C}$ (moreover, a bandlimited function, see Section 1.8). Similarly for $J = [c, d]$ we have

$$\hat{h}_J = -\frac{1}{\mathrm{j}\xi}\left[e^{-\mathrm{j}d\xi} - e^{-\mathrm{j}c\xi}\right],$$

$$\int \hat{h}_I \overline{\hat{h}_j} d\xi = \int_{-\infty}^{\infty} \left[e^{\mathrm{j}(d-b)\xi} - e^{\mathrm{j}(d-a)\xi} - e^{\mathrm{j}(c-b)\xi} + e^{\mathrm{j}(c-a)\xi}\right] \frac{\mathrm{d}\xi}{(2\pi\xi)^2}.$$

There is no pole at the point $\xi = 0$ and we can integrate over the line $\xi = \eta - \imath$. The first term in the bracket gives zero after integration, if $d - b < 0$, since the function $\exp(\mathrm{j}(d-b)\xi)$ decreases rapidly in the bottom half-plane. Otherwise the integral of this term is equal to $\mathrm{j}^2 (d-b)$ (apply the residue theory). Therefore the result of integration equals

$$\int \hat{h}_I \overline{\hat{h}_j} \mathrm{d}\xi = -(d-b)_+ + (d-a)_+ + (c-b)_+ - (c-a)_+,$$

where we set $e_+ = e$ if $e > 0$ and $e_+ = 0$ otherwise. It is easy to see that the right side coincides with the length of the interval $I \cap J$ which is equal to $\int h_I h_J \mathrm{d}x$. This proves (1.21) for $f = h_I, g = h_J$ and for arbitrary step functions f and g.

Take an arbitrary $f \in L_1 \cap L_2$ and an arbitrary $t > 0$; set $f_t(x) = f(x)$ for $|x| \le t$ and $f_t(x) = 0$ otherwise. We have $\|f - f_t\|_1 \to 0$ and $\|f - f_t\|_2 \to 0$ as $t \to \infty$. By Theorem 1.12 we can choose a function h_t that is equal to a linear combination of indicator functions of some intervals $[a, b] \subset [-t, t]$ such that $\|f_t - h_t\|_2 \le 1/t$. We have $\|f_t - h_t\|_1 \le (2t)^{1/2}/t = (2/t)^{1/2}$ by the triangle inequality, consequently $\|f - h_t\|_1 \to 0$ as $t \to \infty$. By 1.4 $\hat{h}_t \to \hat{f}$ uniformly. On the other hand, $\|f - h_t\|_2 \to 0$, which yields that $\{h_t\}$ is a Cauchy sequence in $L_2(\mathbb{R})$. By (1.20) $\left\{\hat{h}_t\right\}$ is a Cauchy sequence in the space $L_2(\mathbb{R}^*)$ and by completeness of this space $\hat{h}_t \to \phi$ in $L_2(\mathbb{R}^*)$. We conclude that $\phi = \hat{f}$, since the same sequence converges to $\hat{f}$ uniformly. Finally

$$\left\|\hat{f}\right\|_2 = \|\phi\|_2 = \lim \left\|\hat{h}_t\right\|_2 = \lim \|h_t\| = \|f\|_2,$$

which proves (1.20). The equation (1.21) follows from the identity $2\,\mathrm{Re}\langle\lambda f, g\rangle = \|\lambda f + g\|^2 - |\lambda|^2\|f\|^2 - \|g\|^2$, where λ is an arbitrary complex number. ▶

Theorem 1.14 (Plancherel) *For an arbitrary function $f \in L_2(\mathbf{V})$ the sequence of functions*

$$F_r(f)(\xi) = \int_{B(r)} \exp(-\mathrm{j}\xi x) f(x) \mathrm{d}x, \ r > 0$$

converges in $L_2(\mathbf{V})$ to a function $\tilde{f}$ as $r \to \infty$ where $B(r)$ means the ball of radius r. The limit satisfies (1.20).

◀ Let again $n = 1$. The integral $F_r(f)$ equals the Fourier transform of the truncation f_r of f. The sequence $\{f_r\}$ converges in L_2 to f as $r \to \infty$, hence it is a Cauchy sequence in L_2. We can apply the Parseval equality, since $f_r \in L_1$: $\|\hat{f}_r - \hat{f}_s\|_2 = \|f_r - f_s\|_2$. It follows that $\hat{f}_r, r = 1, 2, \ldots$ is a Cauchy sequence L_2. Denote by $\tilde{f}$ its limit. We have $\|f_r - f\| \to 0$, which yields $\|\tilde{f}\|^2 = \|f\|^2$. ▶

We call the function $\tilde{f}$ the *Fourier-Plancherel* transform of f. This transform is a continuation of the Fourier transformation, since the function $\tilde{f}$ coincides with (1.4) if $f \in L_1 \cap L_2$. The continuation is unique and keeps the Parseval equality. We use henceforth the same notation $F : f \mapsto \hat{f}$ for the Fourier-Plancherel transform. The Parseval equality means that this operator is unitary as an operator in L_2. The adjoint Fourier transform F^* possesses similar properties and can be extended to the space $L_2(\mathbb{R}^*)$.

Proposition 1.15 *For any $\varphi \in L_2(\mathbb{R})$, $\psi \in L_2(\mathbb{R}^*)$ we have the equation*

$$\langle F(\varphi), \psi \rangle = \langle \varphi, F^*(\psi) \rangle. \tag{1.22}$$

◀ First assume that both functions φ, ψ are integrable and change the order of integration as follows

$$\langle F(\varphi), \psi \rangle = \int \int \exp(-\mathrm{j} \langle \xi, x \rangle) \varphi(x) \mathrm{d}x \bar{\psi}(\xi) \mathrm{d}\xi$$

$$= \int \int \varphi(x) \overline{\exp(\mathrm{j} \langle \xi, x \rangle) \psi(\xi)} \mathrm{d}\xi \mathrm{d}x = \langle \varphi, F^*(\psi) \rangle.$$

For arbitrary square-integrable functions φ, ψ we can apply this equation to the truncated functions φ_r, ψ_r. Then we pass on to the limit as $r \to \infty$ and get (1.22). ▶

Theorem 1.16 *The operators F, F^* are mutually inverse, i.e.,*

$$F^* F = \mathbf{I}, \quad F F^* = \mathbf{I},$$

where $\mathbf{I}$ means the identity operator in $L_2(\mathbf{V})$ and in $L_2(\mathbf{V}^*)$.

◀ The second equation is similar to the first one. The composition F^*F is an isometry. Therefore it is sufficient to prove the first equation on a dense subset of L_2. For indicator functions we can change the order of integration and apply Parseval's theorem:

$$\langle F^* F(h_I), h_J \rangle = \langle F(h_I), F(h_J) \rangle = \langle h_I, h_J \rangle.$$

This implies that $F^*F(h_I) = h_I$ and by continuity of the operator F^*F this equation holds for any function $f \in L_2(\mathbb{R}^*)$. ▶

CONVOLUTION. The integral

$$(f * g)(x) = \int f(x-y)g(y)\mathrm{d}y \tag{1.23}$$

is called *convolution* of integrable functions f, g. The mapping $(f, g) \mapsto f * g$ is a bilinear, commutative and associative operation. It satisfies

$$|f * g| \leq \|f\|_2 \|g\|_2 \tag{1.24}$$

which follows from the triangle inequality.

Proposition 1.17 *If $f, g \in L_1$ the integral* (1.23) *converges almost everywhere and*

$$\|f * g\|_1 \leq \|f\|_1 \|g\|_1.$$

◀ By Fubini's theorem and by changing the variables $z = x - y$ we get

$$\int \left| \int f(y)g(x-y)\mathrm{d}y \right| \mathrm{d}x \leq \int |f(y)||g(x-y)|\mathrm{d}x\mathrm{d}y = \int |f(y)|\mathrm{d}y \int |g(z)|\mathrm{d}z.$$

▶

Theorem 1.18 *For arbitrary $f, g \in L_1$ we have*

$$F(f * g) = F(f)F(g). \tag{1.25}$$

If moreover, $f, g \in L_2$ the symmetric equation is also valid:

$$F(fg) = F(f) * F(g). \tag{1.26}$$

These equations hold for the conjugated transform F^* as well.

◀ First we prove (1.25):

$$\begin{aligned}
F(f * g)(\xi) &= \int \exp(-\mathrm{j}\,\langle \xi, x \rangle) \int f(x-y)g(y)\mathrm{d}y\mathrm{d}x \\
&= \int \exp(-\mathrm{j}\,\langle \xi, z \rangle) \int \exp(-\mathrm{j}\,\langle \xi, y \rangle) f(z)g(y)\mathrm{d}y\mathrm{d}z \\
&= \int \exp(-\mathrm{j}\,\langle \xi, z \rangle) f(z)\mathrm{d}z \int \exp(-\mathrm{j}\,\langle \xi, y \rangle) g(y)\mathrm{d}y.
\end{aligned}$$

The coordinate change $z = x - y$ is eligible, since the integrand belongs to the Lebesgue space $L_1(\mathbf{V} \times \mathbf{V})$. The right side is equal to $F(f)F(g)$. To prove (1.26) we apply (1.25) to the adjoint Fourier transforms of the truncated functions $F(f)_r, F(g)_r$. These functions are integrable since they are bounded; this yields

$$F^*(F(f)_r * F(g)_r) = F^*(F(f)_r)\, F^*(F(g)_r). \tag{1.27}$$

By Plancherel's theorem the sequence $F^*(F(f)_r)$ converges in mean to $F^*(F(f))$ as $r \to \infty$ and $F^*(F(f)) = f$ by Theorem 1.16. Similarly $F^*(F(g)_r) \to g$ in mean. The right side of (1.27) converges to the function fg in the space L_1 by the triangle inequality. Apply the Fourier transform to both sides and get uniform convergence of the sequence $FF^*(F(f)_r * F(g)_r) \to f\,g$. On the other hand the convolution $F(f)_r * F(g)_r$ has compact support and is bounded by virtue of (1.24). It belongs to L_2 hence $FF^*(F(f)_r * F(g)_r) = F(f)_r * F(g)_r$ by Theorem 1.16. By (1.24) $F(f)_r * F(g)_r$ converges uniformly to $F(f) * F(g)$ which implies (1.26). ▶

CONVOLUTION OF DISTRIBUTIONS. The convolution can be correctly defined for some pairs of distributions u, v in $\mathbf{V}$. Let ϕ be a test function; one set $u * \phi(x) = u(\psi_x)$, where $\psi_x(\cdot) = \phi(x - \cdot)$. The function $u * \phi$ is always smooth but need not have compact support. In general the support of the convolution $u * \phi$ is contained in the sum of supports of ϕ and u. The formula

$$u * v(\phi) = u(v * \phi)$$

makes sense if v has compact support. The convolution $u * v$ is well defined as a distribution in $\mathbf{V}$. If u has compact support, the convolution $u*v$ can be defined by a similar formula. If both u, v have compact support, the two definitions coincide. Another option is: $\mathbf{V} = \mathbb{R}$ and the supports of u and v are contained in a ray $\mathbb{R}_+ + a$ for some real a. Then the support of $u * v$ is contained in $\mathbb{R}_+ + 2a$.

Example 22. We have $\operatorname{supp} H^\lambda \subset \mathbb{R}_+$ for the Euler kernels, hence the convolution of kernels H^λ and H^μ is always defined.

Proposition 1.19 $H^\lambda * H^\mu = H^{\lambda+\mu}$ *for* $\lambda, \mu \in \mathbb{C}$.

◀ For $\operatorname{Re}\lambda > 0, \operatorname{Re}\mu > 0$ this is just the Euler identity for Gamma and Beta functions. It remains true for all λ, μ, due to uniqueness of analytic continuation. ▶

Example 23. Equation (1.25) holds also for some pairs of distributions f, g such that the convolution $f * g$ is well defined. In particular, formula (1.5) implies that (1.25) holds for the distribution $f = [\mathrm{d}x/x]$ and any density $g \in L_2(\mathbb{R})$.

POISSON FORMULA

Theorem 1.20 *Let f be a function in $\mathbb{R}$ fulfilling the inequalities*

$$|f(x)| \leq C(|x|+1)^{-a} \quad \text{and} \quad \left|\hat{f}(\xi)\right| \leq C(|\xi|+1)^{-a}$$

for some $a > 1, C > 0$. Then the equation holds,

$$\sum_{\mathbb{Z}} \hat{f}(k) = \sum_{\mathbb{Z}} f(k). \tag{1.28}$$

◀ Note that the inequality for f implies absolute convergence of the right side of (1.28) and integrability of f on the whole line. Therefore the Fourier transform of f is well defined. The inequality for $\hat{f}$ implies absolute convergence of the left side and continuity of f. The series

$$g(x) = \sum_{k\in\mathbb{Z}} f(x+k) \tag{1.29}$$

also absolutely converges and the sum g is a continuous function in $[0,1]$. Take the Fourier series

$$g(x) = \sum_{n\in\mathbb{Z}} g_n \exp(\mathrm{j}nx), \; g_n \doteq \int_0^1 g(x)\exp(-\mathrm{j}nx)\,\mathrm{d}x = \int_{\mathbb{R}} f(x)\exp(-\mathrm{j}nx)\,\mathrm{d}x.$$

Substitute the series (1.29) in the integral and find that $g_n = \hat{f}(n)$. This implies that the series $\sum |g_n|$ converges and the first equation (1.29) holds at each point. Taking $x = 0$ yields (1.28). ▶

Remark. The conditions $f, x^2 f,\ f'' \in L_2(\mathbb{R})$ also imply (1.28).

1.8 Bandlimited functions and interpolation

Theorem 1.21 (Paley-Wiener) *If the support of a function $f \in L_2(\mathbb{R})$ is contained in the interval $[a,b]$ for some a, b, then its Fourier transform $\varphi \doteq F(f)$ has analytic continuation in the complex plane that satisfies the inequality*

$$|\varphi(\zeta)| \le C\exp(2\pi\max(b\eta, a\eta)), \; \zeta = \xi + \imath\eta. \tag{1.30}$$

If a function $\varphi \in L_2(\mathbb{R})$ has a holomorphic continuation in $\mathbb{C}$ that fulfils (1.30), *then the function $f \doteq F^*(\varphi)$ vanishes almost everywhere in $\mathbb{R}\setminus[a,b]$.*

◀ The kernel of the Fourier integral

$$\varphi(\xi) = \int_{-a}^{a} \exp(-\mathrm{j}\xi x) f(x)\,\mathrm{d}x$$

has holomorphic continuation $\exp(-\mathrm{j}\zeta x)$ which is bounded in the segment $[-a,a]$. We have $|\exp(\mathrm{j}\zeta x)| = \exp(2\pi\eta x) \le \exp(2\pi\max(b\eta, a\eta))$ which proves the first statement.

Suppose that φ fulfils (1.30) and $f = F^*(\varphi)$. We show now that $\langle f, g\rangle = 0$ for an arbitrary function $g \in D(\mathbb{R})$ with compact support $\operatorname{supp} g \subset [c,d]$ where $c > b$ or $d < a$. This will imply the second statement. By Parseval's theorem, $\langle f, g\rangle = \langle \varphi, \psi\rangle$ where $\psi = F(g)$. By the first statement the function ψ has analytic continuation in the complex plane which fulfils an estimate like (1.30). Suppose that $c > b$ and consider the function $\psi^*(\zeta) = \bar{\psi}(\bar{\zeta})$; it is analytic too and fulfils

$$|\psi(\zeta)| \le C(|\zeta|+1)^{-2}\exp(-2\pi c\eta), \; \eta > 0.$$

By Cauchy's theorem we can move the chain of integration:

$$\langle \varphi, \psi \rangle = \int \varphi(\xi) \bar{\psi}(\xi) \, d\xi = \int \varphi(\xi + \imath\eta) \psi^* (\xi + \imath\eta) \, d\xi.$$

According to (1.30) we have for an arbitrary $\eta > 0$,

$$|\varphi(\zeta) \psi^* (\zeta)| \leq C \left(|\zeta|^2 + 1 \right) \exp(2\pi (b\eta - c\eta)),$$

which implies $|\langle \varphi, \psi \rangle| \leq C \exp(2\pi (b - c) \eta)$. Taking $\eta \to \infty$, yields $\langle \varphi, \psi \rangle = 0$, q.e.d. In the case $d < a$ we take $\eta \to -\infty$. ▶

Example 24. Consider the function $\psi \doteq (1 - x^2)_+^{\lambda - 1/2} / \Gamma(\lambda + 1/2)$, where $\lambda \in \mathbb{C}$. It belongs to $L_1(\mathbb{R})$ for $\operatorname{Re} \lambda > -1/2$. Take the Fourier transform

$$\hat{\psi} = \pi^\lambda \int_{-1}^{1} \frac{(1 - x^2)^\lambda}{\Gamma(\lambda + 1)} \exp(-\mathrm{j}\xi x) \, \mathrm{d}x = |\xi|^{-\lambda} J_\lambda(\xi).$$

The factor J_λ is called the Bessel function of order λ. By Example 4, the left-hand side has analytic continuation at $\mathbb{C}$, hence the Bessel function $\lambda \mapsto J_\lambda$ has analytic continuation too.

Definition. A function $f \in L_2(\mathbb{R})$ is called an *a-bandlimited* function for some number $a > 0$, if $\operatorname{supp} \hat{f} \subset [-a, a]$.

Example 25. The function $\operatorname{sinc} x = \sin x / x$ is a-bandlimited with $a = 1/2\pi$.

Properties:

1. If $g \in L_2$ and $\operatorname{supp} \hat{g} \subset [b, c]$ for some $b < c$, then the function $f(x) \doteq \exp(-(b + c)\pi\imath x) g(x)$ is a-bandlimited for $a = (c - b)/2$.
2. If f is an a-bandlimited function and g is a b-bandlimited function, then fg is an $a + b$-bandlimited function.
3. If f is an a-bandlimited function, then shift $T_h f$ is again an a-bandlimited function. The derivative f' is a-bandlimited too and $\|f'\|_2 \leq 2\pi a \|f\|_2$. For a proof we take in account the equation $\hat{f'} = \mathrm{j}\xi \hat{f}$. From this $|\hat{f'}| \leq 2\pi a |\hat{f}|$, since $\hat{f}$ vanishes for $|\xi| > a$ and by Parseval's theorem

$$\|f'\|_2^2 = \int |\hat{f'}|^2 \mathrm{d}\xi \leq (2\pi a)^2 \int |\hat{f}|^2 \mathrm{d}\xi = (2\pi a)^2 \|f\|_2^2.$$

4. Any bandlimited function can be interpolated in an arbitrary finite interval from its complement:

Proposition 1.22 *For arbitrary positive a and b any a-bandlimited function $\phi \in L_2(\mathbb{R})$ can be interpolated in $(-b, b)$ as*

$$\phi(\zeta) = \exp\left(2\pi a \sqrt{b^2 - \zeta^2}\right) \int_b^\infty - \int_{-\infty}^{-b} \frac{\sin\left(2\pi a \sqrt{y^2 - b^2}\right)}{\pi (y - \zeta)} \phi(y) dy \qquad (1.31)$$

where the branch is defined by $\operatorname{Re} \sqrt{b^2 - \zeta^2} > 0$.

◀ Denote $\Gamma_+ = [b, \infty), \Gamma_- = (-\infty, b], \Gamma = \Gamma_+ \cup (-\Gamma_-)$. Take the meromorphic form

$$\alpha(z) = \frac{\exp\left(-2\pi a\sqrt{b^2 - z^2}\right)\phi(z)\,\mathrm{d}z}{\jmath(z - \zeta)}$$

defined in $\mathbb{C}\backslash\Gamma$. It tends to zero at infinity due to the Paley-Wiener theorem and is absolutely integrable on each side of Γ. Take the ε-neighborhood $\Gamma_\pm(\varepsilon)$ of $\Gamma_\pm$ and the chain $\gamma(\varepsilon) \doteq \partial\Gamma_+(\varepsilon) - \partial\Gamma_-(\varepsilon)$. By the residue theorem for a point $\zeta \in \mathbb{C}\backslash\Gamma_+(\varepsilon) \cup \Gamma_-(\varepsilon)$,

$$\exp\left(-\pi\sqrt{b^2 - \zeta^2}\right)\phi(\zeta) = -\int_{\gamma(\varepsilon)} \alpha.$$

Choose now $\zeta \in \Gamma$. For two close points $\zeta_\pm \in \gamma(\varepsilon)$ such that $\pm\operatorname{Im}\zeta_\pm > 0$, $\operatorname{Re}\zeta_\pm > 0$, we have

$$\exp\left(-\pi\sqrt{b^2 - \zeta_+^2}\right) - \exp\left(-\pi\sqrt{b^2 - \zeta_-^2}\right) \approx 2\imath\sin\left(\pi\sqrt{\zeta^2 - b^2}\right),$$

where $2\zeta = \zeta_+ + \zeta_-$.

In the case $\operatorname{Re}\zeta_\pm < 0$ we get the quantity $\approx -2\imath\sin\left(\pi\sqrt{\zeta^2 - b^2}\right)$ instead. Therefore

$$-\int_{\gamma(\varepsilon)} \alpha \to \int_\Gamma \frac{\sin\left(\pi\sqrt{z^2 - b^2}\right)}{\pi(z - \zeta)}\phi(z)\mathrm{d}z$$

as $\varepsilon \to 0$. ▶

Corollary 1.23 *If ϕ is an a-bandlimited function, then for any $b > 0$ we have*

$$\int_{-b}^{b} \left|\exp\left(-2\pi a\sqrt{b^2 - x^2}\right)\phi(x)\right|^2 \mathrm{d}x \le \int_{|x|\ge b} |\phi(x)|^2\,\mathrm{d}x.$$

◀ Write the equation (1.31) as

$$\begin{aligned}\exp\left(-2\pi a\sqrt{b^2 - x^2}\right)\phi(x) &= \int_\Gamma \frac{\sin\left(\pi\sqrt{y^2 - b^2}\right)}{\pi(y - x)}\phi(y)\,\mathrm{d}y \\ &= \int \frac{\psi(y)\,\mathrm{d}y}{\pi(x - y)},\ |x| \le b,\end{aligned}$$

where $\psi(y) = \mp\sin\left(\pi\sqrt{y^2 - b^2}\right)\phi(y)$ for $y \in \Gamma_\pm$ and $\psi(y) = 0$ otherwise. The right-hand side equals the Hilbert operator applied to g. The Hilbert operator is unitary, which yields

$$\int_{-b}^{b} \left|\exp\left(-2\pi a\sqrt{b^2 - x^2}\right)\phi(x)\right|^2 \mathrm{d}x = \int |\psi|^2\,\mathrm{d}x \le \int |\phi|^2\,\mathrm{d}x.$$ ▶

Chapter 2

Radon Transform

2.1 Properties

Let $\mathbf{E}$ be a Euclidean space with the interior product $(x,y) \mapsto \langle x,y\rangle$. Take a hyperplane $H \subset \mathbf{E}$, choose a unit orthogonal vector ω to H and denote by p the distance from the origin to H in the direction ω, i.e., $\langle \omega, x\rangle = p$ is the equation of $H = H(p,\omega)$. At the same time $-\langle \omega, x\rangle = -p$ is another equation of the same hyperplane: $H(-p,-\omega) = H(p,\omega)$. Thus we have two-fold covering $\mathbf{S}^{n-1} \times \mathbb{R} \to A_{n-1}(\mathbf{E})$ where $\mathbf{S}^{n-1}$ is the unit sphere in $\mathbf{E}$ and $A_{n-1}(\mathbf{E})$ is the manifold of all hyperplanes in $\mathbf{E}$. The topological space $A_{n-1}(\mathbf{E})$ is homeomorphic to the projective space of dimension n without one point. This point corresponds to the infinite hyperplane in the projective closure of $\mathbf{E}$.

The Euclidean structure in $\mathbf{E}$ generates the volume density (measure) $\mathrm{d}V$ as well as the area density $\mathrm{d}S$ on any hyperplane H in $\mathbf{V}$. Take an arbitrary integrable function f in $\mathbf{E}$ and define the *Radon* transform

$$Rf(H) \doteq \int_H f\mathrm{d}S, \quad Rf(p,\omega) = \int_{H(\omega,p)} f\mathrm{d}S.$$

This integral is well defined for any ω and for almost all $p \in \mathbb{R}$. The function Rf is even, which means that $Rf(-p,-\omega) = Rf(p,\omega)$.

Proposition 2.1 *For an arbitrary $f \in L_1(\mathbf{E})$ the equation holds,*

$$F_{x\to\xi}(f)(\sigma\omega) = F_{p\to\sigma} Rf(p,\omega).$$

This statement is called the "slice theorem".

◀ For a proof we use Fubini's theorem:

$$F_{p\to\sigma} Rf(p,\omega) = \int \exp(-\mathrm{j}\sigma p)\left(\int_{\langle\omega,x\rangle=p} f\mathrm{d}S\right)\mathrm{d}p = \int \exp(-\mathrm{j}\sigma\omega x) f\,\mathrm{d}S\mathrm{d}p.$$

The product $\mathrm{d}S\mathrm{d}p$ is equal to the Euclidean density in $\mathbf{E}$ if we take an Euclidean coordinate system of the form $(y_1 = \langle\omega,x\rangle, y_2, \ldots, y_n)$. This density is invariant with respect to all orthogonal transformations of $\mathbf{E}$, hence $\mathrm{d}S\mathrm{d}p = \mathrm{d}x$ and the right side is equal to

$$\int \exp(-\mathrm{j}\sigma\omega x) f \mathrm{d}x = F(f)(\sigma\omega). \qquad \blacktriangleright$$

Corollary 2.2 *The inversion of the Radon transform can be implemented by inverting of the Fourier transform:* $f = F^*_{\sigma\omega\to x}(F_{p\to\sigma} Rf(p,\omega))$.

BACKPROJECTION OPERATOR This is the operator that transforms an even function $g = g(\omega, p)$ defined on $\mathbf{S}^{n-1} \times \mathbb{R}$ to the function

$$R^* g(x) \doteq \frac{1}{2} \int_{S^{n-1}} g(\langle\omega, x\rangle, \omega) \mathrm{d}\omega.$$

We put here coefficient $1/2$ since the contributions of the opposite points ω and $-\omega$ are equal. In other words the function $g(\langle\omega, x\rangle, \omega)$ is well defined on the projective space $\mathbb{P}^{n-1} \doteq \mathbf{S}^{n-1}/\mathbb{Z}_2$ and $R^* g(x)$ is the integral over this space.

Proposition 2.3 *The equation holds for any continuous integrable function* f,

$$R^* Rf(x) = \frac{\pi^{\frac{n-1}{2}}}{\Gamma\left(\frac{n-1}{2}\right)} \int \frac{f(y)}{|x-y|} \mathrm{d}y.$$

◀ We can assume that $x = 0$ since both sides commute with translations in $\mathbf{V}$.

$$\begin{aligned} 2R^* Rf(0) &= \int_{\mathbf{S}^{n-1}} Rf(0,\omega)\,\mathrm{d}\omega \\ &= \int_{\mathbf{S}^{n-1}} \int_{\langle\omega, y\rangle = 0} f(y)\,\mathrm{d}S\mathrm{d}\omega = \int_{\mathbf{S}^{n-1}} \delta_{\langle\omega,y\rangle}(f)\,\mathrm{d}\omega, \end{aligned} \tag{2.1}$$

where $\mathrm{d}S$ is the area element in the plane $\langle\omega, y\rangle = 0$. The Euler kernels $H^\lambda_\phi = \phi_+^{\lambda-1}/\Gamma(\lambda)$ (see Section 1.6) tend to the Dirac kernel δ_ϕ as $\lambda \to 0$. Take $\phi = \langle\omega, y\rangle$ and evaluate the integral

$$\begin{aligned} \int_{\mathbf{S}^{n-1}} H^\lambda_{\langle\omega,y\rangle}(f)\,\mathrm{d}\omega &= \frac{1}{\Gamma(\lambda)} \int_{\mathbf{S}^{n-1}} \int_{\mathbf{V}} \langle\omega, y\rangle_+^{\lambda-1} f(y)\,\mathrm{d}y\,\mathrm{d}\omega \\ &= \frac{1}{\Gamma(\lambda)} \int_{\mathbf{V}} \int_{\mathbf{S}^{n-1}} \langle\omega, z\rangle_+^{\lambda-1}\,\mathrm{d}\omega\; |y|^{-1} f(y)\,\mathrm{d}y \end{aligned}$$

where we set $z \doteq |y|^{-1} y$. Apply Fubini's theorem to the inner integral:

$$\int_{\mathbf{S}^{n-1}} \langle\omega, z\rangle_+^{\lambda-1}\,\mathrm{d}\omega = \int_0^1 s^{\lambda-1}\mathrm{d}s \int_{\langle\omega,z\rangle = s} \frac{\mathrm{d}\omega}{\mathrm{d}s}.$$

We have $d\omega/ds = (1-s^2)^{n/2-1}\,dv$ where dv is the area element in the unit sphere $\mathbf{S}^{n-2}$. Therefore this integral is equal to

$$\left|\mathbf{S}^{n-2}\right| \int_0^1 s^{\lambda-1}\left(1-s^2\right)^{n/2-1} ds = \frac{\Gamma(\lambda/2)\,\Gamma(n/2)}{2\Gamma((\lambda+n)/2)} \left|\mathbf{S}^{n-2}\right|$$

and

$$\int_{\mathbf{S}^{n-1}} H^{\lambda}_{\langle\omega,y\rangle}(f)\,d\omega = \frac{\Gamma(\lambda/2)\,\Gamma(n/2)}{2\Gamma(\lambda)\,\Gamma((\lambda+n)/2)} \left|\mathbf{S}^{n-2}\right| \int |y|^{-1} f(y)\,dy.$$

The left side tends to (2.1) as $\lambda \to 0$ and the first factor on the right side tends to 1. This completes the proof since $\left|\mathbf{S}^{n-2}\right| = 2\pi^{(n-1)/2}/\Gamma((n-1)/2)$. ▶

2.2 Inversion formulae

For n even we do the substitution $p = \langle\omega, x\rangle + q$ in (2.6) and take into account that $g^{(n-1)}(p,\omega) = \partial_p^{n-1} g(p,\omega)$ is an odd function in p,

$$\begin{aligned} \mathrm{j}^n f(x) &= \lim_{\varepsilon\to 0} \int_{S^{n-1}} \int_{|q|\geq\varepsilon} \frac{g^{(n-1)}(\langle\omega,x\rangle + q,\omega)dq}{q} d\omega \\ &= \lim \int\int_{q\geq\varepsilon} \frac{g^{(n-1)}(\langle\omega,x\rangle + q,\omega) - g^{(n-1)}(-\langle\omega,x\rangle + q,\omega)}{q} dq. \end{aligned}$$

The limit exists if $g^{(n-1)}(p,\omega)$ is a Lipschitz function with respect to p. Now we change the order of integration and write the right side as

$$\int_0^\infty \frac{dq}{q} \left[\int_{S^{n-1}} \left[g^{(n-1)}(\langle\omega,x\rangle + q,\omega) - g^{(n-1)}(-\langle\omega,x\rangle + q,\omega)\right] d\omega\right].$$

Make the substitution $\omega \mapsto -\omega$ in the second integral and see that it gives the same quantity as the first one. Therefore we obtain

$$f(x) = \frac{(-1)^{n/2}}{2^{n-2}\pi^{n/2}\Gamma(n/2)} \int_0^\infty F^{(n-1)}(q)\frac{dq}{q}, \tag{2.2}$$

where

$$F(q) \doteq \frac{1}{|S^{n-1}|} \int g(\langle\omega,x\rangle + q,\omega)d\omega$$

is the normalized back projection and $|S^{n-1}| = 2\pi^{n/2}/\Gamma(n/2)$ is the area of the unit sphere.

Theorem 2.4 *If $f \in S(\mathbf{E})$ and $g = Rf$, then*

$$f(x) = \frac{(n-1)!}{(-\mathrm{j})^n} \int_{S^{n-1}} \int \frac{Rf(p,\omega)dp}{(\langle\omega,x\rangle - p + 0\imath)^n} d\omega. \tag{2.3}$$

◀ According to Proposition 2.1 we can reconstruct the function by means of two Fourier transforms:

$$Rf \mapsto F_{p\to\sigma}Rf \mapsto F^*_{\sigma\omega\to x}\left(F_{p\to\sigma}Rf\right) = f.$$

Write down the right-hand side

$$F^*_{\sigma\omega\to x}\left(F_{p\to\sigma}Rf\right) = \int \exp(\mathrm{j}\sigma\left\langle x,\omega\right\rangle) \int \exp(-\mathrm{j}\sigma p) g(p,\omega)\mathrm{d}p\mathrm{d}\xi, \tag{2.4}$$

where we have set $\xi = \sigma\omega$ and $g = Rf$. We pass to the spherical coordinates in the exterior integral. Apply the equation $\mathrm{d}\xi = \sigma^{n-1}\mathrm{d}\sigma\mathrm{d}\omega$ and change the order of integration:

$$f(x) \stackrel{?}{=} \int_0^\infty \exp(\mathrm{j}\sigma(\langle x,\omega\rangle - p))\sigma^{n-1}\mathrm{d}\sigma \int g(p,\omega)\mathrm{d}p\mathrm{d}\omega.$$

The question mark ? means that the change of variables is not legitimate since the integration with respect to σ obviously diverges. To make it correct we introduce in (2.4) the increasing factor $\exp(-\varepsilon\sigma)$ and write

$$\begin{aligned} f &= \lim_{\varepsilon\to 0} \int \exp(\mathrm{j}x\sigma\omega - \varepsilon\sigma) \int \exp(-\mathrm{j}\sigma p) g(p,\omega)\mathrm{d}p\mathrm{d}\xi \\ &= \int \exp(-\varepsilon\sigma + \mathrm{j}\sigma(\langle x,\omega\rangle - p))\sigma^{n-1}\mathrm{d}\sigma \int\int g(p,\omega)\mathrm{d}p\mathrm{d}\omega. \end{aligned}$$

Now our treatment is legitimate and we need only to evaluate the interior integral

$$I(z) \doteq \int_0^\infty \exp(-z\sigma)\sigma^{n-1}\mathrm{d}\sigma,\ z \doteq \varepsilon - \mathrm{j}(\langle x,\omega\rangle - p).$$

The integrand $\exp(-z\sigma)\sigma^{n-1}\mathrm{d}\sigma$ is the trace on the chain of integration $\mathbb{R}_+ = \{\sigma;\sigma > 0\}$ of a holomorphic form which decreases in the upper half-plane $\mathbb{C}_+ \doteq \{\operatorname{Re}\sigma > 0\}$. By Cauchy's theorem we can replace this chain by the chain $\{\sigma = z^{-1}t,\ t > 0\}$:

$$I(z) = z^{-n}\int_0^\infty \exp(-t)t^{n-1}\mathrm{d}t = z^{-n}\Gamma(n) = \frac{(n-1)!}{(-\mathrm{j})^n(\langle x,\omega\rangle - p + \varepsilon'\imath)^n},$$

where $\varepsilon' \doteq (2\pi)^{-1}\varepsilon$. Therefore

$$f(x) = \frac{(n-1)!}{(-\mathrm{j})^n}\int \lim_{\varepsilon\to 0}\int \frac{g(p,\omega)\mathrm{d}p}{(\langle x,\omega\rangle - p + \varepsilon'\imath)^n}\mathrm{d}\omega.$$

The interior limit is equal to the distribution $(\langle x,\omega\rangle - p + 0\imath)^{-n}$. ▶

Taking the real part of (2.3) we get the expansion of the Dirac function in *plane waves:*

Corollary 2.5 *For any $n \geq 1$,*

$$\delta_0(x) = \frac{(n-1)!}{2\mathrm{j}^n} \int_{\mathbf{S}^{n-1}} \left[\frac{1}{\langle \omega, x - 0\imath \rangle^n} + \frac{(-1)^n}{\langle \omega, x + 0\imath \rangle^n} \right] d\omega. \tag{2.5}$$

Now we obtain inversion formulas in the real form:

Corollary 2.6 *For an arbitrary integrable function f that satisfies the above conditions and $g \doteq Rf$ we have*

$$f(x) = \frac{(-1)^{n/2-1}}{(2\pi)^n} \int_{\mathbf{S}^{n-1}} \int \frac{g^{(n-1)}(\langle \omega, x \rangle - p, \omega)\mathrm{d}p}{p} \mathrm{d}\omega \tag{2.6}$$

for even n (the principal value of the inner integral is taken) and

$$f(x) = \frac{(-1)^{(n-1)/2}}{2(2\pi)^{n-1}} \int_{\mathbf{S}^{n-1}} g^{(n-1)}(\langle \omega, x \rangle, \omega)\mathrm{d}\omega \tag{2.7}$$

for odd n.

Remark. The integrand is an even function of ω, therefore we can remove the coefficient $1/2$, replacing the sphere $\mathbf{S}$ by an arbitrary hemisphere $\mathbf{S}_+$.

◀ We may assume that the function f is real-valued. Then g is real-valued too and we can replace the kernel in (2.3) by its complex conjugate. Taking the sum, we get

$$2f(x) = \frac{(n-1)!}{(-\mathrm{j})^n} \int_{S^{n-1}} \int Rf(p,\omega) \left[(q + 0\imath)^{-n} + (-1)^n (q - 0\imath)^{-n} \right] \mathrm{d}q\mathrm{d}\omega, \tag{2.8}$$

where we set $q \doteq \langle \omega, x \rangle - p$. Calculate the kernel

$$Q_n(q) \doteq (q + 0\imath)^{-n} + (-1)^n (q - 0\imath)^{-n}.$$

Integrating by parts $n-1$ times yields

$$\begin{aligned} \int Q_n(q)a(q)\mathrm{d}q &= \lim_{\varepsilon \to 0} \int \left[(q + \varepsilon\imath)^{-n} + (-1)^n (q - \varepsilon\imath)^{-n} \right] a(q)\mathrm{d}q. \\ &= \frac{1}{(n-1)!} \lim \int \left[(q + \varepsilon\imath)^{-1} + (-1)^n (q - \varepsilon\imath)^{-1} \right] a^{(n-1)}(q)\mathrm{d}q. \end{aligned}$$

By (1.1) we have for n odd,

$$\begin{aligned} &\lim \int \left[(q + \varepsilon\imath)^{-1} + (-1)^n (q - \varepsilon\imath)^{-1} \right] a^{(n-1)}(q)\mathrm{d}q \\ &= \left[(q + 0\imath)^{-1} - (q - 0\imath)^{-1} \right] (a^{(n-1)}) = -\mathrm{j}\delta_0(a^{(n-1)}) = -\mathrm{j}a^{(n-1)}(0). \end{aligned}$$

Apply this equation for $a(q) \doteq Rf(q,\omega)$ and substitute this into (2.8). This gives (2.7). For n even we have by (1.1)

$$\lim \int \left[(q + \varepsilon\imath)^{-1} + (q - \varepsilon\imath)^{-1} \right] a^{(n-1)}(q)\mathrm{d}q = 2\left[q^{-1} \right] (a^{(n-1)}).$$

This together with (2.8) implies (2.6). ▶

Remark. The formulae (2.7) are *local*, i.e., for reconstruction of the value of f at a point x we only need to know the values of $R^{(n-1)}f$ for hyperplanes $H(p,\omega)$ through the point x. Whereas the formulas (2.6) are *non-local*, i.e., we need to know $R^{(n-1)}f$ for *all* hyperplanes.

Now we write both formulas (2.6) and (2.7) in a uniform way by means of the Hilbert operator $\mathbb{H}$.

For any n the following reconstruction formula holds,

$$f = R^* \left(\frac{\mathbb{H}}{2\pi} \frac{\partial}{\partial p} \right)^{n-1} Rf. \tag{2.9}$$

This formula coincides with (2.6) and (2.7) if we take into account that the operators $\mathbb{H}$ and $\partial/\partial p$ commute, $\mathbb{H}^q = (-1)^{q/2} I$ for any even q and the function $(-\mathbb{H})^{n-1} R^{(n-1)} f$ is even in $\mathbf{S}^{n-1} \times \mathbb{R}$.

2.3 Alternative formulae

For even n we can rewrite (2.6) as

$$\begin{aligned}(2\pi)^n f(x) &= -(-1)^{n/2} \int_{S^{n-1}} \int \frac{g^{(n-1)}(p,\omega)\mathrm{d}p}{\langle \omega, x\rangle - p} \mathrm{d}\omega \\ &= (-1)^{n/2} \int \frac{g^{(n-1)}(\langle \omega, x\rangle + q, \omega)\mathrm{d}q}{q} \mathrm{d}\omega,\end{aligned}$$

where the substitution $p = \langle \omega, x\rangle + q$ is applied. We have

$$\begin{aligned}\int \frac{g^{(n-1)}(\langle \omega, x\rangle + q, \omega)\mathrm{d}q}{q} &= \lim_{\varepsilon\to 0} \left(\int_{q\geq\varepsilon} + \int_{q\leq -\varepsilon} \right) \\ = \lim \int_{q\geq\varepsilon} & \frac{g^{(n-1)}(\langle \omega, x\rangle + q, \omega) - g^{(n-1)}(-\langle \omega, x\rangle + q, -\omega)}{q} \mathrm{d}q,\end{aligned}$$

since $g^{(n-1)}(\langle \omega, x\rangle - q, \omega) = -g^{(n-1)}(-\langle \omega, x\rangle + q, -\omega)$ because $g^{(n-1)}(p,\omega)$ is an odd function. The limit exists if $g^{(n-1)}(p,\omega)$ is a Lipschitz function with respect to p. Now we change the order of integration and we get

$$\begin{aligned}&(2\pi)^n f(x) \\ &= (-1)^{n/2} \int_0^\infty \frac{\mathrm{d}q}{q} \int \left[g^{(n-1)}(\langle \omega, x\rangle + q, \omega) - g^{(n-1)}(-\langle \omega, x\rangle + q, -\omega) \right] \mathrm{d}\omega.\end{aligned}$$

Make the substitution $\omega \mapsto -\omega$ in the second integral and see that it gives the same quantity as the first one. Therefore we obtain

$$f(x) = \frac{4\,(-1)^{n/2}}{(2\pi)^n} \int_0^\infty F^{(n-1)}(q) \frac{\mathrm{d}q}{q}, \tag{2.10}$$

where $F(q) \doteq \frac{1}{2} \int g(\langle \omega, x\rangle + q, \omega)\mathrm{d}\omega$ is the backprojection of g.

For odd n we can commute the backprojection operator and derivatives in (2.9):

Proposition 2.7 *A reconstruction can be done by*

$$f = \left(\frac{-\Delta}{4\pi^2}\right)^{(n-1)/2} R^*Rf,$$

where the power of $-\Delta$ is the pseudodifferential operator

$$\left(\frac{-\Delta}{4\pi^2}\right)^p g = F^*\left(|\xi|^{2p} F(g)(\xi)\right). \tag{2.11}$$

◀ By Proposition 2.3 we have $R^*Rf = K * f$, where $K(x) = \frac{\pi^{\frac{n-1}{2}}}{\Gamma\left(\frac{n-1}{2}\right)} |x|^{-1}$. Theorem 1.18 yields $F(R^*Rf) = F(K)\,F(f)$. We have $F(K) = |\xi|^{1-n}$ by Proposition 1.6. This together with (2.11) implies Proposition 2.7. ▶

Specify the above formulae for small dimensions.

2D CASE. (2.9) coincides with Radon's original formula

$$f(x) = -\frac{1}{\pi}\int_0^\infty \frac{\mathrm{d}F(q)}{q}. \tag{2.12}$$

Alternative forms of this equation are:

$$\begin{aligned} f(x,y) &= -\frac{1}{2\pi^2}\int_0^\pi \mathrm{d}\phi \int_{-\infty}^{\infty} \frac{g'(p,\phi)\,\mathrm{d}p}{\cos\phi\, x + \sin\phi\, y - p} \\ &= \frac{1}{2\pi^2}\int_0^\pi \mathrm{d}\phi \int_0^\infty \frac{[g(q+p,\phi) + g(q-p,\phi) - 2g(q,\phi)]\,\mathrm{d}p}{p^2}, \end{aligned} \tag{2.13}$$

where $q = \cos\phi\, x + \sin\phi\, y$.

3D CASE. From (2.7) we find

$$f(x) = -\frac{1}{8\pi^2}\int_{S^{n-1}} g''(\langle\omega, x\rangle, \omega)\mathrm{d}\omega = -\frac{1}{8\pi^2}\Delta\int_{S^{n-1}} g(\langle\omega, x\rangle, \omega)\mathrm{d}\omega. \tag{2.14}$$

Replacing integration by summation yields another reconstruction method:

Theorem 2.8 *Let f be an arbitrary function in $\mathbb{R}^n, n > 1$ supported by the cube $[0,1]^n$, such that $D^i f \in L_2(\mathbb{R}^n)$ for $|i| \le n/2 + 1$. If $\int f\mathrm{d}x = 0$ it can be reconstructed by means of the infinite series*

$$f(x) = \sum_{k\in\mathbb{Z}_0} \frac{1}{|k|} \sum_{q\in\mathbb{Z}} Rf\left(\frac{k}{|k|}, \frac{\langle k, x\rangle + q}{|k|}\right), \tag{2.15}$$

where $\mathbb{G} \subset \mathbb{Z}^n$ is the set of vectors $(k_1, \ldots, k_n)$ such that $\gcd(k_1, \ldots, k_n) = 1$ *and $k_1 \ge 0$.*

Remark. Only a countable number of directions $\omega = |k|^{-1} k$ is used for this reconstruction called "discrete".

◀ From the smoothness assumptions follows that the Fourier series absolutely converges:

$$f(x) = \sum_{i \in \mathbb{Z}^n} \hat{f}_i \exp(\mathrm{j} \langle i, x \rangle), \ \hat{f}_0 = 0.$$

Write $i = qk$ where $k \in \mathbb{Z}_0$ and $q \in \mathbb{Z}$. The number q is uniquely defined for $i \neq 0$ which yields

$$\begin{aligned} f(x) &= \sum_{k \in \mathbb{G}} \sum_q \hat{f}_i \exp(\mathrm{j}q \langle k, x \rangle) = \sum_{k \in \mathbb{G}} \sum_q \int f(y) \exp(\mathrm{j}q \langle k, x - y \rangle) \, \mathrm{d}y \\ &= \sum_{k \in \mathbb{G}} \int \int_{\langle k, y - x \rangle = s} f(y) \frac{\mathrm{d}y}{\mathrm{d} \langle k, y \rangle} \sum_q \exp(-\mathrm{j}qs) \, \mathrm{d}s \\ &= \sum_{k \in \mathbb{G}} \frac{1}{|k|} \int Rf \left(\frac{k}{|k|}, \frac{\langle k, x \rangle + s}{|k|} \right) \sum_q \exp(-\mathrm{j}qs) \, \mathrm{d}s. \end{aligned}$$

The function $h(s) \doteq Rf\left(|k|^{-1} k, |k|^{-1} (\langle k, x \rangle + s)\right)$ fulfils the hypothesis of Poisson's summation formula (Section 1.2). Therefore we can replace the q-sum by the sum of Dirac functions $\delta(s - q)$ taken over all integers q:

$$f(x) = \sum_{k \in \mathbb{G}} \frac{1}{|k|} \sum_q \int Rf \left(\frac{k}{|k|}, \frac{\langle k, x \rangle + q}{|k|} \right).$$ ▶

Remark. Glue the square $[0,1]^2$ to 2-torus T. The sum

$$\sum_{q \in \mathbb{Z}} Rf \left(\frac{k}{|k|}, \frac{\langle k, x \rangle + q}{|k|} \right)$$

can be thought of as the integral over the line Γ_k in T that is the image of the straight line $\{y : \langle k, x - y \rangle + q = 0\}$ in $\mathbb{R}^2$. This is a geodesic of length $|k|$ in the Euclidean metric in the torus. Therefore each term in (2.15) is the average of the function f over a closed geodesic and the right side is the sum over all such means.

In the general case the discrete reconstruction formula of [96] looks as follows:

Theorem 2.9 *Under the same assumption on f the equation holds*

$$f(x) = \hat{f}_0 + \sum_{k \in \mathbb{G}} \left[\frac{1}{|k|} \sum_{q \in \mathbb{Z}} Rf \left(\frac{k}{|k|}, \frac{\langle k, x \rangle + q}{|k|} \right) - \hat{f}_0 \right],$$

where $\hat{f}_0 = \int f \mathrm{d}x$.

A proof can be done along the same lines.

2.4 Range conditions

Let $\mathbf{S}^{n-1}$ be the unit sphere in Euclidean space $\mathbf{E}^n$. It is a smooth manifold: take a hyperplane H in $\mathbf{V}$ and consider the orthogonal projection $p_H : \mathbf{S}^{n-1} \to H$. The mapping p_H is a local coordinate system for any point $s \in \mathbf{S}^{n-1}$ where the tangent plane is not orthogonal to H. If ϕ is a function on the sphere we can extend it to the function $\Phi(x) = \phi(\omega)$ that is defined in $\mathbf{V}\backslash\{0\}$ and is constant in any ray $x = r\omega,\ r > 0$. Apply the Laplace operator Δ to it and take restriction to the sphere:

$$\Delta_{\mathbf{S}}\phi \doteq \Delta\Phi|\mathbf{S}^{n-1},$$

where Δ_S is the spherical Laplace operator. A function h on the sphere is a C^∞-function if and only if the function $\Delta_S^j h$ is well defined and continuous for any j.

Denote by $S(\mathbf{S}^{n-1} \times \mathbb{R})$ the Schwartz space of functions in the product $\mathbf{S}^{n-1} \times \mathbb{R}$, i.e., the space of C^∞-functions $h = h(p, \omega)$ in $\mathbf{S}^{n-1} \times \mathbb{R}$ such that $|p^k \Delta_S^j h^{(i)}(p,\omega)|$ is bounded for any i, j, k. It is easy to check that $Rf \in S(\mathbf{S}^{n-1} \times \mathbb{R})$ if $f \in S(\mathbb{R}^n)$. We show now that the range of the Radon operator $R : S(\mathbb{R}^n) \to S(\mathbf{S}^{n-1} \times \mathbb{R})$ is far from filling the whole space $S(\mathbf{S}^{n-1} \times \mathbb{R})$. First we note that by Fubini's theorem

$$\int Rf(p,\omega)\mathrm{d}p = \int f(x)\mathrm{d}x,$$

hence this integral does not depend on ω! This equation contains a continuum consistency condition. The image of the Radon operator satisfies many more conditions:

Proposition 2.10 *If $f \in S(\mathbb{R}^n)$, then for arbitrary integer $k \geq 0$ the function*

$$m_k(\omega) \doteq \int p^k Rf(p,\omega)\mathrm{d}p \tag{2.16}$$

is equal to the restriction to $\mathbf{S}^{n-1}$ of a homogeneous polynomial of degree k.

◀ We have by Fubini's theorem

$$\int p^k Rf(p,\omega)\mathrm{d}p = \int p^k \left(\int_{H(p,\omega)} f(x)\mathrm{d}S \right) \mathrm{d}p = \int (\langle \omega, x\rangle)^k f(x)\mathrm{d}x.$$

The function $\xi \mapsto \langle \xi, x\rangle^k$ is a homogeneous polynomial of degree ξ, hence the integral of the function $\langle \xi, x\rangle^k f(x)$ is a polynomial of degree k too. The restriction of this polynomial to a unit sphere coincides with (2.16). ▶

This is, in fact, a complete set of consistency conditions:

Theorem 2.11 *Let $g \in S(\mathbf{S}^{n-1} \times \mathbb{R})$ be any function such that for any $k = 0, 1, 2, \ldots$ the moment*

$$m_k(\omega) \doteq \int p^k g(p,\omega)\mathrm{d}p$$

is equal to the restriction of a homogeneous polynomial of degree k. Then there exists a function $f \in S(\mathbb{R}^n)$ such that $Rf = g$.

◀ Consider the integral

$$\phi(\xi) \doteq \int_{-\infty}^{\infty} \exp(-\mathrm{j}|\xi|p) g\left(p, \frac{\xi}{|\xi|}\right) \mathrm{d}p.$$

The function $f \doteq F^* \phi$ will solve the equation $Rf = g$ if we show that $f \in S(\mathbf{V})$. For this we need to show that ϕ is in the Schwartz space $S(\mathbf{V}')$. The crucial point is to check that ϕ is smooth at the origin. Choose a number k and write

$$\exp(-\mathrm{j}|\xi|p) = 1 + \mathrm{j}|\xi|p + \frac{\mathrm{j}^2}{2}\left(|\xi|p\right)^2 + \cdots + \frac{\mathrm{j}^k}{k!}\left(|\xi|p\right)^k + r_k\left(|\xi|p\right),$$

where r_k is a smooth function that vanishes at the origin with its derivatives up to the order k. Substitute in the above integral:

$$\phi(\xi) = \sum_0^k \frac{\mathrm{j}^i}{i!}|\xi|^k \int p^i g\left(p, \frac{\xi}{|\xi|}\right) \mathrm{d}p + \int r_k(|\xi|p) g\left(p, \frac{\xi}{|\xi|}\right) \mathrm{d}p. \tag{2.17}$$

The ith term in the sum is equal to $c_i|\xi|^i m_i(|\xi|^{-1}\xi)$. This is a homogeneous polynomial of degree i of the variables ξ. Therefore the first term in (2.17) is a polynomial of order $\leq k$. We can take any ξ-derivatives of the second term up to the order k. Therefore the sum is, at least, in $C^k(\mathbf{V}^*)$. The function ϕ belongs to $C^\infty(\mathbf{V}^*)$ since the number k is arbitrary. It is easy to check that the all derivatives of ϕ are fast decreasing, hence ϕ is in the Schwartz space. ▶

Remark. It follows that the moment conditions imply that the function g is even.

2.5 Frequency analysis

The range condition of the Radon transform is related to the property of its harmonic decomposition. The Radon transform Rf of a function f with compact support in a Euclidean space $\mathbf{E}^n$ is an even function on the manifold $\Lambda = \mathbb{R} \times \mathbf{S}^{n-1}$. Its harmonic decomposition is the combination of the Fourier decomposition on the first factor and of the expansion into spherical functions on $\mathbf{S}^{n-1}$.

2D CASE. We show that the Fourier density $\hat{g}$ decreases rapidly out of a cone Ξ in the dual group $\hat{E}$ that is equal to $\mathbb{R} \times \mathbb{Z}$ for the parallel beam geometry. This means that restriction of $\hat{g}$ to Ξ is informative enough for a reconstruction of f.

First we find an upper bound for Fourier coefficients of the plane Radon transform in $\mathbb{R}^2$:

$$\hat{g}(\sigma, m) = \frac{1}{4\pi^2}\int_0^{2\pi} \mathrm{d}\varphi \int_{\mathbb{R}} \exp(-\mathrm{j}(\sigma s + m\varphi)) g(s, \varphi) \mathrm{d}s.$$

Theorem 2.12 *For any bounded function f in $\mathbf{E}^2$ with support in the unit disc B and any positive number ε, the Fourier coefficients satisfy the inequality*

$$|\hat{g}(\sigma, m)| \leq \frac{1}{2\pi} \exp(2\pi\left(\sinh(\varepsilon)|\sigma| - \varepsilon|m|\right)) \int |f| \,\mathrm{d}V. \tag{2.18}$$

◀ We have

$$\hat{g}(\sigma, m) = \int\int\int_{p=s} \exp(-\mathrm{j}(\sigma p + m\varphi)) f \mathrm{d}q\, \mathrm{d}p\, \mathrm{d}\varphi,$$

where $p = \cos\varphi\, x + \sin\varphi\, y$, $q = -\sin\varphi\, x + \cos\varphi\, y$. Changing variables we get

$$\hat{g}(\sigma, m) = \int_B f(x,y) \left[\int_{-\pi}^{\pi} \exp(-\mathrm{j}\Phi(x,y;\varphi))\, \mathrm{d}\varphi \right] \mathrm{d}x\, \mathrm{d}y, \tag{2.19}$$

where $\Phi = \sigma(\cos\varphi\, x + \sin\varphi\, y) + m\varphi$. The phase function Φ admits a holomorphic continuation in φ in the complex plane $\mathbb{C}$, whereas $\exp(-\mathrm{j}\Phi)$ has analytic continuation in the cylinder $\mathbb{C}/2\pi\mathbb{Z}$. Denote by $\zeta \doteq \varphi + \imath\psi \pmod{2\pi}$ the complex coordinate in the cylinder. We have $\operatorname{Im}\Phi = \sigma q \sinh\psi + m\psi$. Move the chain of integration in (2.19) to the circle $\psi = -\operatorname{sign}(m)\varepsilon$ for some $\varepsilon > 0$. By Cauchy's theorem we get the same quantity

$$\int_{-\pi}^{\pi} \exp(-\mathrm{j}\Phi(x,y;\varphi))\, \mathrm{d}\varphi = \int_{-\pi}^{\pi} \exp(-\mathrm{j}\Phi(x,y;\varphi + \imath\psi))\, \mathrm{d}, \varphi$$

where

$$|\exp(-\mathrm{j}\Phi(x,y;\varphi+\imath\psi))| = \exp(\operatorname{Im}\Phi(x,y;\varphi+\imath\psi)) = \exp(\sigma q \sinh\psi + m\psi)$$

and $|\sigma q \sinh\psi + m\psi| \leq -|m|\varepsilon + \sigma\sinh\varepsilon$ for $(x,y) \in B$. Therefore

$$\left| \int_{-\pi}^{\pi} \exp(-\mathrm{j}\Phi(x,y;\varphi+\imath\psi))\, \mathrm{d}\varphi \right| \leq 2\pi \exp(-|m|\varepsilon + \sigma\sinh\varepsilon).$$

By (2.19) this implies (2.18). ▶

3D CASE. In this case we have the decomposition

$$g(s,\omega) = \sum_{lm} \int \hat{g}(\sigma, l, m) \exp(-\imath\sigma s)\, \mathrm{d}s\, Y_{lm}(\omega),$$

where Y_{lm} are spherical functions. We show now that the harmonic density $\hat{g}(\sigma, l, m)$ is essentially supported in the cone $\Sigma \doteq \{\sigma \geq l\}$ where l is the number of the corresponding irreducible representation of the group $O(3)$. We may say that the quantities σ, l, m are related like quantum numbers of electrons in Bohr's atom $\sigma > l \geq |m|$ if we restrict ourselves to an essential support of the density $\hat{g}$.

Theorem 2.13 *For any bounded function f in $\mathbf{E}^3$ whose support is contained in the unit ball, the Fourier coefficients $\hat{g}$ of Rf satisfy for any positive ε, k the equation*

$$|\hat{g}(\sigma, l, m)| \leq C_k \left(\frac{|\sigma|}{l(l+1) - \sigma^2} \right)^{-k} \tag{2.20}$$

as $\sigma^2 \to \infty, \quad 1 \leq \sigma^2 \leq (1-\varepsilon) l(l+1).$

◀ The spherical harmonics form an orthogonal basis on the sphere. Therefore

$$\hat{g}(\sigma,l,m)=\int_{s^2}\int_{\mathbb{R}} g(s,\omega)Y_{lm}(\omega)\exp(-\imath\sigma s)\,\mathrm{d}s\,\mathrm{d}\omega,$$

where $\mathrm{d}\omega=\sin\theta\mathrm{d}\theta\wedge\mathrm{d}\varphi$ is the Euclidean density on the sphere. Hence

$$\hat{g}(\sigma,l,m)=\int\int\int_{p=s} f\,Y_{lm}(\theta,\varphi)\exp(-\imath\sigma s)\,\mathrm{d}S\,\mathrm{d}s\,\mathrm{d}\omega,$$

where $\mathrm{d}S$ is the Euclidean density in the plane $p=s$ where $p\doteq\langle\omega,x\rangle$. The product $\mathrm{d}S\,\mathrm{d}s$ is equal to the Euclidean measure on $\mathbf{E}$, hence

$$\hat{g}(\sigma,l,m)=\int_B I_{lm}f\,\mathrm{d}x,\ \text{where}\quad I_{lm}\doteq\int Y_{lm}(\theta,\varphi)\exp(-\imath\sigma p)\,\mathrm{d}\omega.$$

Now we estimate the integral I_{lm}. The spherical functions satisfy the equation $(\Delta+L)Y_{lm}=0$ where $L=l(l+1)$ and

$$\Delta=\frac{1}{\sin\theta}\frac{\partial}{\partial\theta}\sin\theta\frac{\partial}{\partial\theta}+\frac{1}{\sin^2\theta}\frac{\partial^2}{\partial\varphi^2}$$

is the Laplace operator on the sphere in the standard coordinates θ,φ where $\omega=(\sin\theta\cos\varphi,\sin\theta\sin\varphi,\cos\theta)$. Define the gradient operator in the sphere $\nabla a=\left(a'_\theta,\,a'_\varphi/\sin\theta\right)$ and take into account the formula $\Delta(ab)=\Delta(a)b+2(\nabla a,\nabla b)+a\Delta b$. For the phase function $\Phi=\sigma p$ we have

$$\Delta\Phi=-2\Phi,\quad |\nabla\Phi|^2=\sigma^2\left(|x|^2-p^2\right). \tag{2.21}$$

Write the identity

$$\int\frac{\exp(-\imath\Phi)\,(\Delta+L)Y_{lm}\mathrm{d}\omega}{L-\sigma^2|x|^2}=0$$

and take the integral by parts. The operator Δ is symmetrical with respect to the form $\mathrm{d}\omega$ and $\Delta\exp(-\imath\Phi)=-\left(|\nabla\Phi|^2+2\Phi\right)\exp(-\imath\Phi)$. Integrate by parts and apply (2.21)

$$0=\int Y_{lm}\frac{(\Delta+L)\exp(-\imath\Phi)}{L-\sigma^2|x|^2}\mathrm{d}\omega=I_{lm}+2\imath\int\frac{Y_{lm}\Phi\exp(-\imath\Phi)\,\mathrm{d}\omega}{L-\sigma^2|x|^2}. \tag{2.22}$$

The last integral admits the estimate $O\left(\sigma/\left(L-\sigma^2\right)\right)$, since $|x|\le 1$ and $\Phi=O(\sigma)$. This proves the assertion of the theorem with $k=1$.

Then we apply similar arguments to all the integrals in the right side of (2.22) and get the estimate $I_{lm}=O\left(\left(\sigma/\left(L-\sigma^2\right)\right)^2\right)$. Repeating this procedure we prove (2.20) for $k=2,3,\ldots$. ▶

2.6 Radon transform of differential forms

Let **E** be an oriented Euclidean space of dimension n and σ be an $n-1$-differential form in **E** whose coefficients decrease at infinity sufficiently fast. The integral

$$I\sigma(H) = \int_H \sigma$$

is defined for an arbitrary hyperplane $H = H(p,\omega)$ in **E**. We define the orientation of H by means of the normal vector ω; thus the orientation of $H(-p,-\omega)$ is the opposite of that of $H(p,\omega)$. We call $I\sigma$ the affine integral transform of the form σ. This transform can be described in terms of vector fields. Choose a system of coordinates $x_1,\ldots,x_n$ in **E** and take the volume form $\mathrm{d}x = \mathrm{d}x_1 \wedge \cdots \wedge \mathrm{d}x_n$. For a vector (tangent) field s, we set $\sigma = s \vdash \mathrm{d}x$. Then the equation

$$I\sigma(H(p,\omega)) = \langle \omega, R[s](p,\omega)\rangle = \sum \omega_i R[s_i](p,\omega)$$

holds, where R denotes the Radon transform, $s = \sum s_i \partial/\partial x_i$ and

$$\sigma = \sum (-1)^{i-1} s_i \mathrm{d}x_1 \wedge \cdots \widehat{\mathrm{d}x_i} \cdots \wedge \mathrm{d}x_n.$$

We have $I\mathrm{d}\tau = 0$, if τ is an $n-2$-form that decreases together with $\mathrm{d}\tau$ sufficiently rapid at infinity. Therefore the form σ cannot be reconstructed from data of hyperplane integrals. However, it is possible to reconstruct its differential.

Proposition 2.14 *If the form* $\mathrm{d}\sigma$ *is integrable, it can be explicitly reconstructed from data of* $I\sigma$.

◀ Take an arbitrary half-space $U \subset \mathbf{V}$; let H be the oriented boundary of U. By the Green formula

$$\int_H \sigma = \int_U \mathrm{d}\sigma.$$

Write $\mathrm{d}\sigma = f\mathrm{d}x$ for an integrable function f. Taking the derivative of the right-hand side in the normal direction ω yields

$$\frac{\partial}{\partial p}\int_{H(p,\omega)} \sigma = \int_{H(p,\omega)} \mathrm{d}\sigma = R[f](p,\omega).$$

Thus we know the Radon transform of f. Apply an inversion formula and we recover f which gives $\mathrm{d}\sigma = f\mathrm{d}x$. ▶

In terms of the vector field s this means that the *divergence*

$$\operatorname{div} s = \sum_i \frac{\partial s_i}{\partial x_i}$$

can be reconstructed. Indeed, for $\sigma = s \vdash \mathrm{d}x$ we have $\mathrm{d}\sigma = \operatorname{div} s\, \mathrm{d}x$.

Chapter 3

The Funk Transform

3.1 Factorable mappings

Let X be a Riemannian manifold of dimension n with the metric tensor g and $\mathsf{Y} = \{Y\}$ be a family of closed subvarieties of X of dimension k, $0 < k \leq n$. For a continuous function f in X that decreases sufficiently fast at infinity, we define the integrals

$$Mf(Y) = \int_Y f \, \mathrm{d}V(Y), \; Y \in \mathsf{Y} \tag{3.1}$$

where $\mathrm{d}V(Y)$ is the volume element on Y induced by the metric g. We call the function $Mf|\mathsf{Y}$ an *integral* transform of f. For a Euclidean space X and the family of hyperplanes we call this operator a Radon transform. We follow this terminology in any situation where the geometry is symmetric with respect to a transitive commutative group; another example: tori.

We call the transformation (3.1) a *Funk* transform, if the manifold X possesses a noncommutative symmetry group or in the case of algebraic varieties Y. For the sphere $X = \mathbf{S}^2$ and the family of big circles C, this transform coincides with the classical Minkowski-Funk transform. Alternatively we can take the projective plane $X = \mathbb{P}^2$ instead of the sphere and the family of projective lines Y, which are images of big circles.

In some special geometrical situations a reconstruction formula for the transform (3.1) can be translated for another geometry by means of the following simple arguments:

Definition. Let $\Phi : X_1 \to X_2$ be a smooth mapping of Riemannian manifolds (it need not to be an isometry), $\mathsf{Y} = \{Y\}$ be a family of smooth subvarieties of X_1. We say that Φ is (infinitesimally) *factorable* with respect to this family if for an arbitrary $Y \in \mathsf{Y}$ and arbitrary point $x \in Y$ the equation holds for the Jacobian of Φ:

$$\frac{\mathrm{d}V_{(2)}\left(\Phi\left(x\right), \Phi\left(Y\right)\right)}{\mathrm{d}V_{(1)}\left(x, Y\right)} = j\left(x\right) J\left(Y\right), \tag{3.2}$$

where $\mathrm{d}V_{(i)}\left(x, Z\right)$ is the Riemannian volume density on a submanifold $Z \subset X_i$, $i = 1, 2$ at a point x. The functions $j : X \to \mathbb{R}$ and $J : \mathsf{Y} \to \mathbb{R}$ depend on Φ only;

we call these functions *Jacobian factors* of the mapping Φ. An application of this property is obvious: the problem of inversion of the integral operator M for the family $\Phi(Y) = \{\Phi(Y), Y \in \mathsf{Y}\}$ is reduced to that for the family Y by

$$Mf(\Phi(Y)) = \int_{\Phi(Y)} f \mathrm{d}V_{(2)} = J(Y) \int_Y \Phi^*(f) j(x) \mathrm{d}V_{(1)}$$
$$= J(Y) M[\Phi^*(f) j](Y),$$

where f is a function on X_2 and $\Phi^*(f)(x) \doteq f(\Phi(x))$. If there is an inversion operator I for $M|\mathsf{Y}$, we define the inversion operator for $M|\Phi(\mathsf{Y})$ as follows: $f = I(JM[\Phi^*(f)j])$. The reduction can be reverted since the mapping Φ^{-1} is also factorable. Moreover, the transitivity property holds:

Proposition 3.1 *If a mapping $\Psi : X_2 \to X_3$ is factorable for the family $\Phi(\mathsf{Y})$ and a Riemannian space X_3, then the composition $\Psi\Phi : X_1 \to X_3$ is factorable for Y with the Jacobian factors*

$$j_{\Psi\Phi}(x) = j_\Psi(\Phi(x)) j_\Phi(x), \quad J_{\Psi\Phi}(Y) = J_\Psi(\Phi(Y)) J_\Phi(Y).$$

Example 1. Any conformal mapping Φ possesses the property (3.2) for the family Y_k of all k-dimensional subvarieties with the Jacobian factors $j = (\Phi^*(g_2)/g_1)^{k/2}$, $J = 1$, where $g_{1,2}$ are the metric tensors in $X_{1,2}$.

Example 2. Let D be the unit disc in the Euclidean plane. The automorphism of D given by $G(z) = 2z\left(1 + |z|^2\right)^{-1}$ is factorable for the family of circle arcs $A \subset D$ that are orthogonal to the circle ∂D. The Jacobian factors are

$$j_G(z) = \frac{1 - |z|^2}{\left(1 + |z|^2\right)^2}, \quad J_G(A) = \frac{2\left(1 + r^2\right)^{1/2}}{r},$$

where r is the radius of an arc A. The image $G(A)$ is the chord in D that props up the arc A.

Example 3. Let $\mathbf{E}$ be a Euclidean space and L be a projective transformation of the projective closure $\bar{\mathbf{E}}$ of $\mathbf{E}$. It defines the mapping $L : \mathbf{E} \backslash L^{-1}(H_\infty) \to \mathbf{E} \backslash L(H_\infty)$ where H_∞ denotes the improper hyperplane in $\bar{\mathbf{E}}$. This mapping is factorable for the variety of k-dimensional affine subspaces $A \subset \mathbf{E}$ and arbitrary k. To show this property, we take the Euclidean space $\mathsf{E} = \mathbb{R} \dotplus \mathbf{E}$ with the coordinates $x_0, \ldots, x_n$ and consider the isometrical embedding $e : \mathbf{E} \to \mathsf{E}$, $e(x_1, \ldots, x_n) = (1, x_1, \ldots, x_n)$. Take a linear automorphism $\mathbf{L}$ of E that generates L, i.e., $L = p\mathbf{L}e$ where $p : \mathsf{E} \backslash \{x_0 = 0\} \to \mathbf{E}$ is the central projection, $p(x) = (x_1/x_0, \ldots, x_n/x_0)$. Set $p_0(x) = x_0$.

For arbitrary points $b^0, \ldots, b^k \in \mathsf{E}$ denote by $\{b^0, \ldots, b^k\}$ the $k \times (n+1)$ matrix with rows $b^0, \ldots, b^k$. The positive number $\left[b^0, \ldots, b^k\right]$ is defined by

$$\left[b^0, \ldots, b^k\right]^2 = \sum_{0 = i_0 < i_1 < \cdots < i_k} \left|B_{i_0, \ldots, i_k}\right|^2,$$

where $B_{i_0,\dots,i_k}$ means the minor of the matrix $\{b^0,\dots,b^k\}$ formed by columns with numbers $i_0,\dots,i_k$. Note that $[a^0,\dots,a^k] = k!V$ for arbitrary points $a^0,\dots,a^k \in E$, where V is the volume of the simplex $(a^0,\dots,a^k)$.

Proposition 3.2 *For an arbitrary affine k-plane A in $\mathbf{E}$ the equation*

$$\frac{\mathrm{d}V(L(x),L(A))}{\mathrm{d}V(x,A)} = |p_0(\mathbf{L}(e(x)))|^{-k-1}\frac{[\mathbf{L}(a^0),\dots,\mathbf{L}(a^k)]}{[a^0,\dots,a^k]} \tag{3.3}$$

holds, where $a^0,\dots,a^k$ are arbitrary points in $\mathbf{E}$ that span A, i.e.,

$$A = \left\{x = \sum t_i a^i, \sum t_i = 1\right\}.$$

◀ The quotient $[\mathbf{L}(a^0),\dots,\mathbf{L}(a^k)]/[a^0,\dots,a^k]$ does not change if we replace $a^0,\dots,a^k$ by another set of vectors that span A. Therefore it is sufficient to check (3.3) for an infinitesimal simplex $(a^0,\dots,a^k)$, i.e., for the case

$$a^0 = x,\ a^i = x + \varepsilon u^i,\ p_0(u^1) = \dots = p_0(u^k) = 0. \tag{3.4}$$

Suppose that the mapping $\mathbf{L}$ preserves the n-plane $e(\mathbf{E})$. Then the quantity $[\mathbf{L}(a^0),\dots,\mathbf{L}(a^k)]$ is equal to $k!W$ where W is the volume of the simplex with vertices $\mathbf{L}(a^0),\dots,\ \mathbf{L}(a^k)$ and the right side of (3.3) is equal to W/V since $p_0(\mathbf{L}(e(x))) = 1$. This proves (3.3) for any $\mathbf{L} = \mathbf{L}_0$ preserving $e(\mathbf{E})$. This is true also for arbitrary matrix $\mathbf{L}$ whose first line is equal to $(l,0,\dots,0)$ since dividing by l does not change the right side of (3.3). Now we check it for any coordinate permutation of the form

$$\mathbf{L}_j(x) = (x_j, x_1,\dots,x_{j-1},x_0,x_{j+1},\dots,x_n).$$

This will imply the proposition since an arbitrary operator $\mathbf{L}$ is a composition of transformations $\mathbf{L}_0,\mathbf{L}_1,\dots,\mathbf{L}_n$ and the right side of (3.3) is multiplicative with respect to composition.

Take $j = 1$ for simplicity of notation. We have

$$L(1,x_1,\dots,x_n) = (1,y_1,\dots,y_n),\ y_1 = 1/x_1, y_2 = x_2/x_1,\dots,y_n = x_n/x_1.$$

Take a simplex of the form (3.4). Its volume is equal to $\varepsilon^k[x,u^1,\dots,u^k]$ (up to the factor $(k!)^{-1}$). By (1.25) the volume of the simplex $(L(a^0),\dots,L(a^k))$ equals $\varepsilon^k[L(x),v^1,\dots,v^k]$ where

$$v^j = x_1^{-2}\left(0,-u_1^j, x_1u_2^j - x_2u_1^j,\dots,x_1u_n^j - x_nu_1^j\right),\ j = 1,\dots,k.$$

It is easy to check that

$$\begin{aligned}\{L(x),v^1,\dots,v^k\} &= x_1^{-1}\begin{pmatrix} y_1 & 1 & y_2 & \dots & y_n \\ u_1^1 & 0 & u_2^1 & \dots & u_n^1 \\ \cdot & \cdot & \cdot & \dots & \cdot \\ u_1^k & 0 & u_2^k & \dots & u_n^k \end{pmatrix} \\ &= x_1^{-1}\{L(x),L(u^1),\dots,L(u^k)\}.\end{aligned}$$

Therefore

$$\frac{\mathrm{d}V\left(L\left(x\right),L\left(A\right)\right)}{\mathrm{d}V\left(x,A\right)}=\frac{\left[L\left(x\right),v^{1},\ldots,v^{k}\right]}{\left[x,u^{1},\ldots,u^{k}\right]}=x_{1}^{-k-1}\frac{\left[L\left(x\right),L\left(u^{1}\right),\ldots,L\left(u^{k}\right)\right]}{\left[x,u^{1},\ldots,u^{k}\right]},$$

which coincides with the right side of (3.4) since $x_1 = p_0\left(L\left(e\left(x\right)\right)\right)$. ▶

Example 4. Let $\mathbf{S}$ be the unit sphere in a Euclidean space $\mathbf{E}$ with the center in the origin, $\mathbf{E}\backslash\{0\}\rightarrow\mathbf{S}$ be the central projection. Take an arbitrary sphere S in $\mathbf{E}\backslash\{0\}$. The projection defines the mapping $\pi : S\rightarrow\mathbf{S}$. It is factorable for the family of k-spheres $F\cap S$ where F is an arbitrary $k+1$-subspace of $\mathbf{E}$. The volume relation is

$$\frac{\mathrm{d}V\left(x,F\cap S\right)}{\mathrm{d}V\left(x_{0},F\cap\mathbf{S}\right)}=\frac{\left|x\right|^{k+1}}{\left|\left\langle x,x-\xi\right\rangle\right|}r_{F},\ x_{0}=\pi\left(x\right), \tag{3.5}$$

where ξ is the center of S, r_F is the radius of the sphere $F\cap S$. The Jacobian factors are $j\left(x\right)=\left|\left\langle x,x-\xi\right\rangle\right|^{-1}\left|x\right|^{k+1}, J\left(F\right)=r_F$. If the origin is inside S, the dominator $\left\langle x,x-\xi\right\rangle$ does not vanish on S; otherwise it vanishes at each point where a ray from the origin is tangent to S.

Example 5. We can take a hyperplane $H\subset\mathbf{E}\backslash\{0\}$ instead of the sphere S in Example 4. The central projection $\pi : H\rightarrow\mathbf{S}$ is again factorable:

$$\frac{\mathrm{d}V\left(x,F\cap H\right)}{\mathrm{d}V\left(x_{0},F\cap\mathbf{S}\right)}=\frac{\left|x\right|^{k+1}}{\operatorname{dist}\left(F\cap H,0\right)},\ x_{0}=\pi\left(x\right). \tag{3.6}$$

This follows from (3.5) if we move ξ to infinity along the line orthogonal to H at x.

Example 6. CORMACK'S CURVES. Take a positive number α and consider the family of curves in a Euclidean plane $\mathbf{E}$,

$$A_{\alpha}\left(p,\theta\right):\ r^{\alpha}\cos\left(\alpha\left(\phi-\theta\right)\right)=p^{\alpha},\ \left|\alpha\right|<\pi/2\alpha,$$

called α-curves. This family is rotation invariant. In the case $\alpha = 1/2, 1, 2$ the curve A_α is a parabola, straight line and one branch of a hyperbola, respectively. For a positive β the β-curve is defined by the equation

$$B_{\beta}\left(p,\theta\right):\ p^{\beta}\cos\left(\beta\left(\theta-\phi\right)\right)=r^{\beta},\ \left|\beta\right|<\pi/2\beta.$$

For $\beta = 1/2, 1, 2$ a β-curve is a cardioid, a circle through the origin and one branch of a Bernoulli lemniscate, respectively. These two types of curves are dual: fixing a point x in the plane, the parameters (p,θ) of the curves A_α through x belong to a curve B_α and vice versa. For a function f in $\mathbf{E}$, Cormack defined the integral transform for α-curves

$$Mf\left(p,\theta\right)=\int_{A\left(p,\theta\right)}f\left(x\right)\mathrm{d}s$$

and similarly for β-curves. He found in [11] the inversion formulae for these integral transforms by means of harmonic decomposition f and Mf with respect to the polar angle.

If α is an integer or inverse integer we can apply an alternative method by reduction to the Radon transform. Let $\alpha = 1/n$; consider the conformal mapping $w = z^n$ defined in $\mathbf{E}\setminus\{0\}$. Any curve $A_{1/n}$ coincides with the image of a straight line L and

$$\int_{A_{1/n}} f(w)\,\mathrm{d}s = \int_L f(z)\,n\,|z|^{n-1}\,\mathrm{d}s.$$

From this equation we know all line integrals of $g(z) = f(z)\,n\,|z|^{n-1}$ and can recover g by inverting the plane Radon transform. In the case $\alpha = n$ the image of A_{2n} is again a straight line L and

$$\int_{A_n} f(z)\,\mathrm{d}s = \int_L f\left(n\,|z|^{n-1}\right)^{-1}\mathrm{d}s.$$

The inversion of the α-transform can be again done by means of Radon's formula. The Cormack transform for β-curves can be inverted in a similar way by applying the conformal mapping $\psi(z) = z^{-n}$, $n = 2, 3, \ldots$.

We consider more examples of factorable mappings in Chapter 8.

3.2 Spaces of constant curvature

There are three types of complete simply connected Riemannian manifolds of constant sectional curvature: elliptic, Euclidean and hyperbolic. A Euclidean space has zero curvature. Any straight line is a geodesic and vice versa. An elliptic space of dimension n is the real projective space $\mathbb{P}^n = \mathbf{S}^n/\mathbb{Z}_2$ with the metric inherited from the unit sphere $\mathbf{S}^n \subset \mathbf{E}$ where $\mathbf{E}$ is a Euclidean space of dimension $n+1$. The sectional curvature of the elliptic space is equal everywhere to 1. For a subspace $F \subset \mathbf{E}$ of dimension two, the intersection $F \cap \mathbf{S}^n$ is a big circle; its image in the elliptic space $\mathbb{P}^n$ is a closed geodesic curve γ. For an arbitrary subspace F the manifold $Y \doteq F \cap \mathbf{S}^n/\mathbb{Z}_2$ is a projective subspace. It is a totally geodesic manifold of $\mathbb{P}^n$, since any two points of Y can be connected by a geodesic $\gamma \subset Y$. Vice versa, any closed totally geodesic submanifold of the projective space is equal to a projective subspace.

A hyperbolic space of dimension n can be constructed in a similar way. Choose Euclidean coordinates $x_0, x_1, \ldots, x_n$ in $\mathbf{E}$ and consider the hyperboloid $\mathbf{Q} \subset \mathbf{E}$ given by

$$x_0^2 = x_1^2 + \cdots + x_n^2 + 1,\ x_0 > 0.$$

This is one sheet of the quadratic hypersurface. The hypersurface $\mathbf{Q}$ endowed with the induced pseudo-Euclidean metric $\mathrm{d}\sigma^2 = -\mathrm{d}s_0^2 + \mathrm{d}x_1^2 + \cdots + \mathrm{d}s_n^2$ is a hyperbolic space H of sectional curvature -1. For any subspace $F \subset \mathbf{E}$ of dimension 2 the intersection $F \cap \mathbf{Q}$ is a closed geodesic curve; for an arbitrary F it is a totally geodesic submanifold of H.

Alongside this, we consider the Euclidean submanifold $E \doteq \{x_0 = 1\}$ and hemisphere $\mathbf{S}_+ \doteq \{x; |x| = 1, x_0 > 0\}$ endowed with the metric $\mathrm{d}s^2 = \mathrm{d}x_0^2 + \cdots +$

ds_n^2. This is a model of the elliptic space $\mathbb{P}^n$. The central projection π in $\mathbf{E}\setminus\{0\}$ defines the algebraic mapping

$$\mathbf{Q} \xrightarrow{\pi} \mathbf{E} \xleftarrow{\pi} \mathbf{S}_+. \tag{3.7}$$

We have $\pi(\mathbf{S}_+) = \mathbf{E}$ and this mapping is invertible. The set $\pi(\mathbf{Q})$ is an open ball of radius 1. By the aforesaid, the intersection of these subvarieties with a subspace F i … ic space, res…

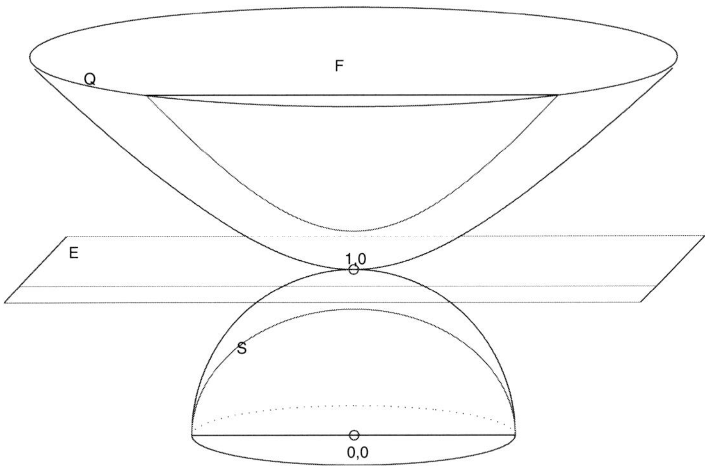

Figure 3.1: Spaces of constant curvature

Example 7. The mappings (3.7) are factorable, namely for an arbitrary subspace $F \subset \mathbf{E}$ of dimension $k+1$ we have

$$\frac{\mathrm{d}V_S(y, F\cap \mathbf{S}_+)}{\mathrm{d}V_E(x, F\cap \mathbf{E})} = \left(1 + d^2(F)\right)^{1/2}\left(1+|x|^2\right)^{-(k+1)/2}, \tag{3.8}$$

$$\frac{\mathrm{d}V_H(y, F\cap \mathbf{Q})}{\mathrm{d}V_E(x, F\cap \mathbf{E})} = \left(1 - d^2(F)\right)^{1/2}\left(1-|x|^2\right)^{-(k+1)/2}, \tag{3.9}$$

where $\pi(y) = x$ is a point in $\mathbf{E}$ where the volume forms are compared and $d(F) = \mathrm{dist}_{\mathbf{E}}(F\cap\mathbf{E}, 0)$. Note that (3.8) is equivalent to (3.5).

For a Euclidean space $\mathbf{E}$ and for a hyperbolic space H and for the sphere $\mathbf{S}$ we consider the integral transform (3.1) on the variety of totally geodesic manifolds of dimension k. In all three cases, (3.1) is called a *geodesic* integral transform. We call the geodesic integral transform in a Euclidean space *flat* integral transform. The flat integral transform coincides with the Radon transform for affine subspaces of dimension $k = n-1$ and with the X-ray transform for $k = 1$.

Corollary 3.3 *The geodesic integral transform in hyperbolic space* H^n *is equivalent to the flat integral transform in the unit ball* $\mathbf{B} \subset \mathbf{E}^n$. *The flat integral transform in* $\mathbf{E}^n$ *is reduced to the Funk transform in the elliptic space* $\mathbf{S}^n_+$.

The inverse reduction does not hold: the Funk transform is reduced to the affine transform only for functions f in $\mathbf{S}_+$ that vanish sufficiently fast on the equator, more precisely, the following condition is necessary: $f(y) = o\left(y_0^{k+1}\right)$ as $y \in \mathbf{S}_+, y_0 \to 0$.

3.3 Inversion of the Funk transform

We write an inversion formula for the integral transform $M_{\mathbf{S}}$ acting on the manifold of big spheres $Y \subset \mathbf{S} = \mathbf{S}^n$ of dimension $n-1$. Let $\mathbf{S}^*$ be the dual sphere; a point $z \in \mathbf{S}^*$ defines the polar set $z^\perp \doteq \{x; \langle x, z\rangle = 0\}$. It is the projective subspace of $\mathbf{S}$ of dimension $n-1$. Let g be a bounded function on $\mathbf{S}$; the function $M_{\mathbf{S}} g\left(z^\perp\right)$ is defined in $\mathbf{S}$. Take the point $y = (1,0) \in \mathbf{S} \subset \mathbf{E}^{n+1}$ and consider the family of spheres $\{z; \langle z, y\rangle = \cos\phi\}, 0 \le \phi \le \pi/2$ in $\mathbf{S}^n$. The average of $M_{\mathbf{S}} g\left(z^\perp\right)$ over the sphere equals

$$G(\phi, y) = \frac{1}{|S^{n-1}|} \int_{\langle z,y\rangle = \cos\phi} M_{n-1} g\left(z^\perp\right) \mathrm{d}S.$$

Theorem 3.4 *If* n *is even, then any sufficiently smooth even function* g *on* $\mathbf{S}^n$ *is reconstructed by*

$$g(y) = \frac{(-1)^{n/2}}{2^{n-2}\pi^{n/2}\Gamma(n/2)} \int_0^{\pi/2} \left(\cos\phi \frac{\partial}{\partial\phi}\cos\phi\right)^{n-1} G(\phi) \frac{\mathrm{d}\phi}{\sin\phi}.$$

◀ By (3.8) $G(\phi, y) = \sec\phi\, F(y, \tan\phi)$ where $F(r)$ is the spherical mean of hyperplane integrals $M_{\mathbf{E}} f(H)$, $f = \left(1 + |x|^2\right)^{-n/2} g$ over the sphere

$$\{d(H) = r = \tan\phi\}.$$

By (2.2) the reconstruction is given by

$$\begin{aligned} g(y) &= c_n \int_0^\infty F^{(n-1)}(r) \frac{\mathrm{d}r}{r} \\ &= c_n \int_0^{\pi/2} \left(\cos\phi \frac{\partial}{\partial\phi}\cos\phi\right)^{n-1} G(\phi) \frac{\mathrm{d}\phi}{\sin\phi}, \end{aligned}$$

where $c_n^{-1} = 2^{n-2}(-\pi)^{n/2}\Gamma(n/2)$. ▶

Theorem 3.5 *If* n *is odd, we have*

$$g(y) = \frac{(-1)^{(n-1)/2}}{2(2\pi)^{n-1}} \int_{S^{n-1}} \left(\cos\phi \frac{\partial}{\partial\phi}\cos\phi\right)^{n-1} M_{\mathbf{S}} g(0, \omega)\, \mathrm{d}\omega.$$

◀ This follows from (3.8) and (2.7). ▶

The integrand can be simplified as follows:

$$\left(\cos\phi\frac{\partial}{\partial\phi}\cos\phi\right)^{n-1}|_{\phi=0}=\left(\frac{\partial}{\partial\phi}\right)^{n-1}+c_{n,2}\left(\frac{\partial}{\partial\phi}\right)^{n-3}+\cdots+c_{n,n-1}$$

for some integers $c_{n,k}, k=2,4,\ldots,n-1$.

3.4 Radon's inversion via Funk's inversion

FUNK'S INVERSION FORMULA. For $n=2$, integrating by parts yields

$$\begin{aligned}-\pi g(y)&=\int_0^{\pi/2}\left(\frac{\cos^2\phi}{\sin\phi}\frac{\partial}{\partial\phi}-\cos\phi\right)G(\phi)\,\mathrm{d}\phi\\&=\int_0^{\pi/2}\left(\frac{\cos^2\phi}{\sin\phi}+\sin\phi\right)G'(\phi)\,\mathrm{d}\phi\\-G\left(\frac{\pi}{2}\right)&=\int_0^{\pi/2}G'(\phi)\frac{\mathrm{d}\phi}{\sin\phi}-G\left(\frac{\pi}{2}\right),\end{aligned}$$

which implies

$$g(y)=-\frac{1}{\pi}\int_0^{\pi/2}\frac{\mathrm{d}G(y,\phi)}{\sin\phi}+\frac{1}{\pi}G\left(y,\frac{\pi}{2}\right). \tag{3.10}$$

Let f be an arbitrary continuous function in $\mathbf{E}^2$ such that $f=o\left(|x|^{-2}\right)$ at infinity. Define the function $g(y)\doteq(1+|x|^2)f(x)$ on $\mathbf{S}_+$ where $\pi(y)=x$. It tends to zero as y approaches the equator of $\mathbf{S}_+$; we set $g=0$ on the equator and extend it to $\mathbf{S}^2$ as an even function. By (3.8) we have, for an arbitrary big circle C,

$$M_{\mathbf{S}}g(C)=\int_C g\mathrm{d}V_{\mathbf{S}}=(1+d^2(L))^{1/2}\int_L f\mathrm{d}V_{\mathbf{E}}=(1+d^2(L))^{1/2}M_{\mathbf{E}}f(L), \tag{3.11}$$

where $L=\pi(C)$ is a straight line in $\mathbf{E}^2$ and $M_{\mathbf{E}}$ is the flat integral transform in the Euclidean space. If we know $M_{\mathbf{E}}f$, we can determine $M_{\mathbf{S}}g$ and apply (3.10) for the point $x=0, y_0=(1,0)$ taking into account that the second term vanishes. The quantity $G(y,\phi)$ is the average of $M_{\mathbf{S}}g(C)$ for big circles whose spherical distance from y is equal to ϕ. We have $1+d^2(L)=\cos^{-2}\phi$ for $L=\pi(C)$ and by (3.11) the right side is equal to

$$\frac{1}{2\pi\cos q}\int_{d(L)=\tan\phi}M_{\mathbf{E}}f(L)\mathrm{d}\phi=\frac{F(x,\tan\phi)}{\cos\phi}$$

where

$$F(x,r)\doteq\frac{1}{2\pi}\int_{d(L)=r}M_{\mathbf{E}}f(L)\,\mathrm{d}\phi. \tag{3.12}$$

Substitute this equation in (3.10) and change the variable ϕ to $r = \tan \phi$:

$$g(y) = -\frac{1}{\pi} \int_0^\infty \frac{\mathrm{d}(\sec \phi\, F(y,r))}{\sin \phi}.$$

We have

$$\frac{\mathrm{d}(\sec \phi\, F(y,r))}{\sin \phi} = \frac{\mathrm{d}F}{r} + r\mathrm{d}F + F\mathrm{d}r = \frac{\mathrm{d}F}{r} + \mathrm{d}(rF).$$

The last term vanishes after integration over the ray $(0, \infty)$ since the product rF vanishes at the ends. The equation $rF = o\left(r^{-1}\right)$ for $r \to \infty$ follows from $g = o\left(|x|^{-2}\right)$. This yields

$$f\,(0) = g(y) = -\frac{1}{\pi} \int_0^\infty \frac{\mathrm{d}F\,(0,r)}{r}.$$

Moving the origin to an arbitrary point $x \in \mathbf{E}$, yields

$$f\,(x) = -\frac{1}{\pi} \int_0^\infty \frac{\mathrm{d}F\,(x,r)}{r}. \tag{3.13}$$

This is Radon's formula, see Section 2.2.

The same arguments applied to the projection $\pi : \mathbf{Q} \to \mathbf{E}$ give by means of (3.9) the inversion formula for the hyperbolic plane

$$g\,(x) = -\frac{1}{\pi} \int_0^\infty \frac{\mathrm{d}G\,(x,q)}{\sinh q}, \tag{3.14}$$

where G is again defined by (4.12) and q is the hyperbolic distance.

Comparing the formulae (2.10),(3.13) and (3.14), we see the obvious similarity. The form of the denominators: $r, \sin r, \sinh r$ shows direct impact of Euclidean, elliptical and hyperbolic geometries, respectively.

3.5 Unified form

Consider a vector space $\mathbf{V} = \mathbf{V}^{n+1}$ with a coordinate system $x = (x_0, \ldots, x_n)$; define the volume density $\mathrm{d}x = \mathrm{d}x_0 \wedge \cdots \wedge \mathrm{d}x_n$. Take the Euler field $\varepsilon = \sum x_j \partial/\partial x_j$ and define the even differential form

$$\omega \doteq \varepsilon \vdash \mathrm{d}x = \sum_0^n (-1)^j\, x_j \mathrm{d}x_0 \wedge \cdots \widehat{\mathrm{d}x_j} \cdots \wedge \mathrm{d}x_n.$$

Let f be a homogeneous function in $\mathbf{V} \backslash \{0\}$ of degree $-n$, i.e., $f\,(tx) = t^{-n} f\,(x)$ for any $x \in \mathbf{E}, x \neq 0$, $t > 0$. Define its Funk-Radon transform (FR-transform) as

$$I(f)\,(\xi) = \frac{1}{2} \int_K f\,(x)\, \delta\,(\langle \xi, x \rangle)\, \omega\,(x)\,, \tag{3.15}$$

where K is a hypersurface in $\mathbf{V} \backslash \{0\}$ that meets each ray with the source at the origin only once and ξ runs over the dual space $\mathbf{V}^*, \xi \neq 0$.

Proposition 3.6 *The integral* (3.15) *does not depend on the hypersurface K that is homotopic in $\mathbf{V}\setminus\{0\}$ to the sphere $\mathbf{S}$ centred at the origin. It is an even homogeneous function in $\mathbf{V}^*\setminus\{0\}$ of degree -1.*

◀ It is sufficient to check that the integrand is a closed form (current). The form $\mathrm{d}x$ is closed and $\mathrm{d}(\varepsilon \vdash \mathrm{d}x) = (n+1)\,\mathrm{d}x$ which yields $\mathrm{d}(f\delta_\xi\omega) = \mathrm{d}(f\delta_\xi) \wedge \omega + f\delta_\xi(n+1)\,\mathrm{d}x$ where $\delta_\xi(x) = \delta(\langle \xi, x\rangle)$. On the other hand, $\varepsilon(\omega) = 0$ and $\varepsilon(f\delta_\xi) = -(n+1)f\delta_\xi$, since the Dirac function δ_ξ is homogeneous of degree -1. This yields $\varepsilon \vdash \mathrm{d}(f\delta_\xi\omega) = \varepsilon(f\delta_\xi)\,\omega + (n+1)f\delta_\xi\,(\varepsilon \vdash \mathrm{d}x) = 0$ which proves the first statement. To check the second one we note that the Dirac function δ_ξ is even and homogeneous of degree -1 with respect to ξ. ▶

Take an arbitrary even function g on the sphere $\mathbf{S}$ and extend it to a function f in $\mathbf{V}\setminus\{0\}$ as a homogeneous function of degree $-n$. Then

$$I[f](\xi) = \frac{1}{2}\int_{\mathbf{S}\cap\xi^\perp} g(y)\,\omega_\xi = \int_{\mathbf{S}_+\cap\xi^\perp} g(y)\,\omega_\xi,$$

where $\xi^\perp$ is the polar set $\{y : \langle \xi, y\rangle = 0\}$ and $\omega_\xi = \omega/\mathrm{d}\langle \xi, y\rangle$. Introduce the Euclidean structure in $\mathbf{V}$ by means of coordinates x, then the restriction of ω to the oriented unit sphere $\mathbf{S}$ coincides with the area density $\mathrm{d}x/\mathrm{d}r$. If ξ is a unit covector, ω_ξ coincides with the Euclidean area density on $\mathbf{S}\cap\xi^\perp$. This implies that I coincides with the Funk transform.

To compare (3.15) with the Radon transform, we take the Euclidean hyperplane $E_1 \doteq \{y : y_0 = 1\}$ in $\mathbf{E}$. If h is a function on E_1, we can extend it to $\mathbf{E}$ as an even homogeneous function of degree $-n$ as follows:

$$f(y) = |y_0|^{-n}\, h(y_1/|y_0|, \dots, y_n/|y_0|).$$

We have $\omega|E_1 = \mathrm{d}x_1 \wedge \cdots \wedge \mathrm{d}x_n$ and $I[f](\xi) = R[h](H)$ where $H = E_1 \cap \xi^\perp$ is a hyperplane in E_1: $H = H(p,\theta)$, where $\theta = \langle \xi, \eta\rangle\eta - \xi$, $p = \langle \xi, \eta\rangle / |\theta|$ and $\eta = (1, 0, \dots, 0)$.

Similarly the geodesic transform on the hyperbolic space is reduced to the transform (3.15) for functions f supported in the cone

$$\left\{ y : y_0 > \left(y_1^2 + \cdots + y_n^2\right)^{1/2} \right\}.$$

Corollary 3.7 *The equations* (3.8) *and* (3.9) *hold.*

Inversion formulae. Let ω^* be the n-form in $\mathbf{E}^*$ defined similarly to ω. Take an arbitrary half-space $\mathbf{E}_+$ in $\mathbf{E}$.

Theorem 3.8 *The operator I maps the space of even homogeneous functions f in $\mathbf{E}$ of degree $-n$ isomorphically to the space of even homogeneous functions g in $\mathbf{E}^*$ of degree -1. The inversion is given by the transformation*

$$J[g](x) = \frac{1}{(2\pi\imath)^{n-1}} \int_{K^*\cap\mathbf{E}_+} g(\xi)\,\delta^{(n-1)}(\langle x, \xi\rangle)\,\omega^*(\xi)$$

for odd n and

$$J[g] = \frac{\imath(n-1)!}{(2\pi\imath)^{n-1}} \int_{K^* \cap \mathbf{E}_+} g(\xi) \langle x, \xi \rangle^{-n} \omega^*(\xi)$$

for even n, where K^* is an arbitrary hypersurface in $\mathbf{E}^* \setminus \{0\}$ that is homotopic to the unit sphere.

3.6 Funk-Radon transform and wave fronts

Definition. [46] Let u be a distribution in an open set $X \subset \mathbf{V}$. The *wave front* $WF(u)$ is the subset in $X \times \mathbf{V}_0^* \setminus \{0\}$ defined as follows: a point (x, ξ), $x \in X, \xi \in \mathbf{V}^*$ does not belong to $WF(u)$ if $\xi = 0$ or there exists a neighborhood U of x and a conic neighborhood V of ξ in $\mathbf{V}^*$ such that for any function $\phi \in D(U)$ the Fourier transform of ϕu fulfils

$$F(\phi u)(\eta) = O\left(|\eta|^{-q}\right), \ \eta \in \mathbf{V}, |\eta| \to \infty, \ q = 0, 1, 2, \dots .$$

The wave front set is a closed conic subset of $X \times \mathbf{V}_0^* \setminus \{0\}$. The wave front of a generalized function v is defined to be $WF(v\mathrm{d}x)$ where $\mathrm{d}x$ is a smooth volume form in $\mathbf{V}$.

This definition is generalized for distributions on a smooth manifold X. The role of the space $X \times \mathbf{V}^*$ is played by the *cotangent* bundle $T^*(X)$ of X (see Chapter 1). The wave front of a distribution $u \in D'(X)$ is a conic subset of $T_0^*(X)$; the bottom note means that only non-zero covectors are collected. A conic set is a union of rays, so one can reduce the cotangent bundle to the set of rays in it.

Definition. An oriented contact element in a manifold X is a set of covectors $\{(x, t\xi), t \in \mathbb{R}_+\}$, i.e., a ray in the cotangent bundle. The manifold of all oriented contact elements is called the contact bundle of X; we denote it by $C^*(X)$. It is isomorphic to $T_0^*(X)/\mathbb{R}_+$ where $\mathbb{R}_+$ is the multiplicative group of positive numbers. For any submanifold $Y \subset X$ there is defined the submanifold $N^*(Y) \subset C^*(X)$, called conormal bundle to Y; it consists of all pairs (x, ω) where $x \in Y$ and ω is orthogonal to Y.

The wave front of a distribution in a vector space $\mathbf{V}$ can be easily defined in terms of the Radon or the Funk transform. The Funk transform helps to do it in a more natural way. We embed the vector space $\mathbf{V}$ to the unit sphere $\mathbf{S}$ in a Euclidean space $\mathbf{E}$ of dimension $\dim \mathbf{V} + 1$ by means of the central projection. Let $\mathbf{S}^*$ be the unit sphere in the dual Euclidean space $\mathbf{E}^*$. We have the duality pairing $\mathbf{S} \times \mathbf{S}^* \to \mathbb{R}$, $(x, \xi) \mapsto \xi x$. Let $\mathbf{F} \subset \mathbf{S} \times \mathbf{S}^*$ be the hypersurface where $\langle \xi, x \rangle = 0$. The manifold $\mathbf{F}$ can serve as the contact bundle $C^*(\mathbf{S})$ over $\mathbf{S}$: for a pair (x, ξ) the vector ξ is interpreted as a unit cotangent vector at x. The same manifold is equivalent to the contact bundle $C^*(\mathbf{S}^*)$, so we have the involution $\mathsf{L} : C^*(\mathbf{S}) \to C^*(\mathbf{S}^*)$, $(x, \xi) \mapsto (\xi, x)$, called the *Legendre* transform.

Proposition 3.9 *For a distribution $u \in D'(U)$, $U \subset \mathbf{S}$, a point $(x,\xi) \in \mathbf{F}$ does not belong to the set $WF(u) \cup -WF(u)$ if there exists a test function ϕ in $\mathbf{S}$, wiht $\phi(x) \neq 0$, such that the function $M[\phi u]$ is smooth in a neighborhood of ξ. Vice versa, if the point (x,ξ) does not belong to this set, then there exist a neighborhood U of x and a neighborhood $\mathbf{V}$ of ξ such that $M[\phi u]$ is smooth in $\mathbf{V}$ for any test function ϕ,* $\operatorname{supp}\phi \subset U$.

The notation $-G$ means the set of points $(x,-\xi)$ for $(x,\xi) \in G$.

◀ We write the Funk transform of a generalized function on the sphere in the form

$$Mv(\xi) = \int_{\xi^{\perp}} v\,\mathrm{d}S,\ \xi \in \mathbf{S}^{*},$$

where $\xi^{\perp}$ is the polar set of a point ξ and $\mathrm{d}S$ is the area form in $\xi^{\perp}$. We can shift to the Radon transform by means of (3.8), which yields

$$Rw(p,\omega) = \left(1 + d^{2}\left(\xi^{\perp}\right)\right)^{1/2} M[\phi v](\xi),$$

where $w = \left(1 + |x|^{2}\right)^{-n/2} \phi v$ and the arguments (p,ω) of the Radon transform relate to the point $\xi = (\xi_0, \xi') \in \mathbf{S}^{*}$ as follows: $p = \xi_0/|\xi'|, \omega = \xi'/|\xi'|$. The function $M[\phi v]$ is smooth in a neighborhood of a point ξ if and only if the Fourier transform of the left side,

$$F(w)(\eta) = \int \exp(-\mathrm{j}pt)\, Rw(p,\omega)\,\mathrm{d}p,$$

decreases rapidly as $\eta \doteq t\omega$ tends to infinity in a conic set that contains a neighborhood of the vector $\omega = \xi'/|\xi'| \in \mathbf{E}$. This means that (x,ω) belongs to $WF[\phi v]$ for no x and vice versa. ▶

Proposition 3.10 *If u is a distribution in $\mathbf{S}$ with compact support, then the set $WF(Mu) \cup -WF(Mu)$ is contained in $\Sigma \doteq \mathsf{L}(WF(u))$.*

Remark. It follows that the singular support Mu is contained in the set $p\mathsf{L}(WF(u))$ where $p : C^{*}(\mathbf{S}^{*}) \to \mathbf{S}^{*}$ is the natural projection. In other words, the singular support of Mu is contained in the set of points $\xi \in \mathbf{S}^{*}$ such that $(x,\xi) \in WF(u)\,\langle \xi, x\rangle$ for some $x \in \mathbf{S}$.

◀ Take a point $(x,\xi) \in \Sigma \backslash WF(u)$ and show that the point (ξ, x) does not belong to $WF(Mu) \cup -WF(Mu)$. According to Proposition 3.9 it is sufficient to find a test function ψ in $\mathbf{S}^{*}$ such that $\psi(\xi) \neq 0$ and the function $M^{\sharp}[\psi Mu]$ is smooth in a neighborhood of x, where $M^{\sharp}$ is the Funk transform of functions on $\mathbf{S}^{*}$:

$$M^{\sharp}[w](x) = \int_{x^{\perp}} w\mathrm{d}S.$$

By Proposition 3.9 there exists a neighborhood U of x such that $M^\sharp [\psi M [\phi u]]$ is smooth in $\mathbf{S}$ for any test function ϕ supported in U and for a test function ψ in $\mathbf{S}^*$ such that $\psi(\xi) \neq 0$. Take for ϕ a function that is equal to 1 in a neighborhood U_1 of x. Then we have

$$M^\sharp [\psi M u] = M^\sharp [\psi M [\phi u]] + M^\sharp [\psi M [(1-\phi) u]] .$$

The first term in the right side is smooth everywhere. The second term is smooth in U_1 since the singularity of the kernel $M^\sharp [\psi M]$is contained in the diagonal in $\mathbf{S} \times \mathbf{S}$ and the distribution $(1-\phi) u$ vanishes in U_1. Therefore the sum $M^\sharp [\psi M u]$ is smooth in U_1. ▶

Example 8. Let χ be the indicator function of a domain $D \subset \mathbf{S}$ and g be a smooth function in $\mathbf{S}$. The Funk transform of $f = g\chi$ is equal to the integral

$$Mf(\xi) = \int_{D \cap \xi^\perp} g \mathrm{d}S.$$

The wave front of f is contained in the cotangent bundle $N^*(\Gamma)$ of the boundary, moreover it coincides with this bundle if g is nowhere flat in Γ.

3.7 Integral transform of boundary discontinuities

Let D be a domain in $\mathbf{E}^n$ with the smooth boundary $\Gamma \doteq \partial D$. Take a smooth function ϕ in $\mathbf{E}^n$ that is negative in D and positive in the complement and $\mathrm{d}\phi \neq 0$ in Γ; set $\phi_\pm = \max(\pm\phi, 0)$. Consider the generalized functions

$$f_\lambda(x) = a(x) \frac{\phi_-^\lambda(x)}{\Gamma(\lambda+1)}, \ \lambda \in \mathbb{C}, \tag{3.16}$$

where $a \in D(\mathbf{E}^n)$ is a smooth function. It is supported in D and smooth except for the boundary ∂D. Moreover, the wave front of f_λ is contained in the conormal bundle $N^*(\partial D)$. According to Section 1.6 the family $\{f_\lambda\}$ has analytic continuation $[f_\lambda]$ in the complex plane $\lambda \in \mathbb{C}$, in particular,

$$[f_{-k-1}](\rho) = \delta_\phi^{(k)}(a\rho), \ k = 0, 1, \ldots, \ \rho \in D(\mathbf{E}),$$

where

$$\delta_\phi^{(k)}(\rho) = \int_{\phi=0} \left(\frac{\mathrm{d}}{\mathrm{d}\phi}\right)^{k-1} \frac{\rho \mathrm{d}x}{\mathrm{d}\phi}$$

is a derivative of the Dirac function on the boundary. The Funk-Radon transform of f_λ is defined in the usual sense for any λ, $\operatorname{Re}\lambda > 0$. It admits analytic continuation too. We shall show that the singularity of this transform has singularity of a similar form on the dual hypersurface. Take embedding $e : \mathbf{E}^n \rightarrow \mathbf{E}^{n+1}, x \mapsto (1, x)$ and the projection $p : \mathbf{E}^{n+1} \rightarrow \mathbf{S}^n$. The composition $pe : \mathbf{E}^n \rightarrow \mathbf{S}^n$ is the standard central projection as in Section 3.2. The Radon transform in $\mathbf{E}$ and the Funk transform in $\mathbf{S}$ are related as follows:

$$R\left[f_{\lambda}\right](p,\omega)=\left(1+d^{2}\left(\xi^{\perp}\right)\right)^{1/2}M\left[\left(1+|x|^{2}\right)^{n/2}f_{\lambda}\right](\xi).$$

They have singularities of the same form. The support of these singularities is the hypersurface $\Gamma^{*}\doteq\mathsf{L}\left(N^{*}\left(pe\left(\Gamma\right)\cap U\right)\right)\subset\mathbf{S}^{*}$, called the *Legendre dual hypersurface* to Γ. The Legendre dual to Γ^{*} coincides with Γ since L is an involution.

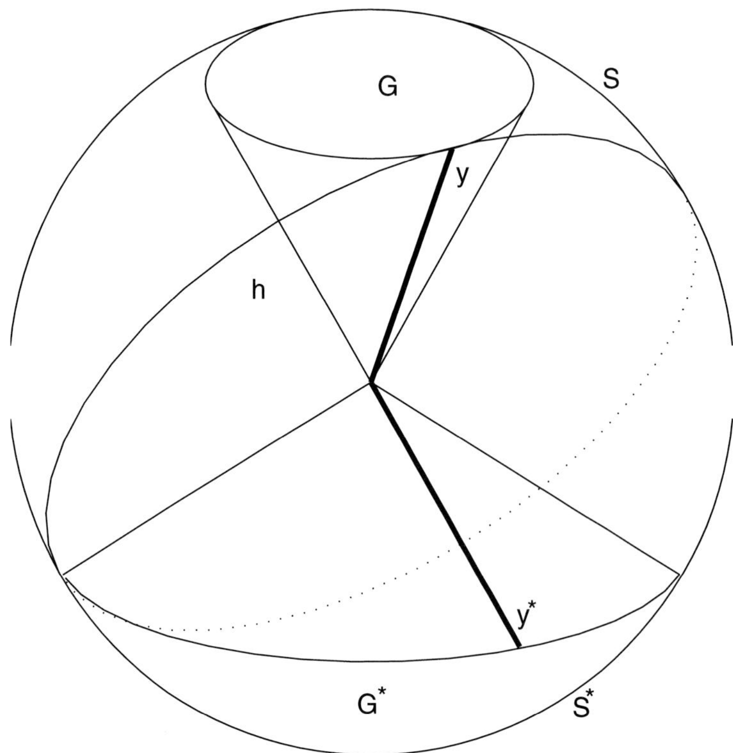

Figure 3.2: Legendre duality in spherical geometry

We consider now the relation of geometries of dual hypersurfaces in Euclidean coordinate systems:

Proposition 3.11 *Let Γ be a smooth hypersurface in $\mathbf{E}^{n}$ such that the Gauss curvature does not vanish at a point $x_0\in\Gamma$. Then the dual hypersurface Γ^{*} is smooth at the point ξ_0 that is an exterior normal to D at x_0 and its Gauss curvature does not vanish. Moreover, the signatures of curvature forms of Γ and of Γ^{*} coincide if we choose properly the side D^{*} of Γ^{*}. In particular, D^{*} is convex if D is convex.*

◄ Choose the coordinates x in $\mathbf{E}^n$ in such a way that $\langle \xi_0, x_0 \rangle > 0$ and replace ξ_0 by $\langle \xi_0, x_0 \rangle^{-1} \xi_0$. Denote by $\mathbf{y} = (t, x)$ the coordinates in $\mathbf{E}^{n+1}$ and by $\xi = (\tau, \xi)$ the coordinates in the dual space $\left(\mathbf{E}^{n+1}\right)^*$. The vector $\xi_0 \doteq (-1, \xi_0)$ is orthogonal to $(1, x_0)$ and to the tangent plane $T(\Gamma)$. Therefore it is orthogonal to the tangent plane $T(G)$ to the cone $G \subset \mathbf{E}^{n+1}$ generated by Γ. Take the function $\Phi(\mathbf{y}) \doteq t^2 \phi(x/|t|)$ in $\mathbf{E}^{n+1}$; it is homogeneous of degree 2 and coincides with ϕ in $e(\mathbf{E}^n)$. Consider the equation

$$\nabla \Phi(\mathbf{y}) = \xi \tag{3.17}$$

for a point $\xi \in \left(\mathbf{E}^{n+1}\right)^*$ close to ξ_0. For $\xi = \xi_0$ it has the solution $\mathbf{y}_0 = (1, x_0)$. Consider the Jacobian $n+1$-matrix of this system at this point:

$$\nabla^2_{t,x} \Phi(\mathbf{y}_0) = \begin{pmatrix} -2\langle x_0, \nabla\phi \rangle + {}^t x_0 \nabla^2 \phi\, x_0 & \nabla' \phi - {}^t x_0 \nabla^2 \phi \\ \nabla \phi - \nabla^2 \phi\, x_0 & \nabla^2 \phi \end{pmatrix},$$

where $\nabla^i \phi = \nabla^i \phi(x_0)$, $i = 1, 2$ and ${}^t x_0$ means the row vector x_0.

Lemma 3.12 *The matrix $\nabla^2 \Phi(\mathbf{y}_0)$ is non-singular.*

◄ Multiply the columns with numbers 2 up to $n = 1$ by the coordinates of vector x_0 and add to the first column. Then multiply the lower n lines by the same numbers and add to the first row. This results in the matrix

$$J \doteq \begin{pmatrix} 0 & \nabla' \phi \\ \nabla \phi & \nabla^2 \phi \end{pmatrix}.$$

It is sufficient to show that the matrix J is non-singular. Suppose that $J(v_0, v) = 0$ for some $n+1$-vector (v_0, v). Taking the first row, we find that $\langle \nabla \phi, v \rangle = 0$ which means that v belongs to the tangent space $T_{x_0}(\Gamma)$. Then $\nabla \phi\, v_0 + \nabla^2 \phi\, v = 0$ and consequently ${}^t u\, \nabla^2 \phi\, v = 0$ for an arbitrary vector $u \in T_{x_0}(\Gamma)$. By the assumption the restriction of the quadratic form $\nabla^2 \phi$ to $T_{x_0}(\Gamma)$ is non-singular, which implies that $v = 0$ and $v_0 = 0$, q.e.d. By the implicit function theorem the equation (3.17) has a unique smooth solution $\mathbf{y} = \mathbf{y}(\xi)$ in a neighborhood $\mathbf{V}$ of the point ξ_0 such that $\mathbf{y}(\xi_0) = \mathbf{y}_0$. We have $\mathbf{y}(\lambda \xi) = \lambda \xi$ for $\lambda > 0$, since $\nabla \Phi$ is homogeneous of degree 1. Taking derivatives of (3.17) yields $\nabla^2 \Phi(\mathbf{y}(\xi))\, \mathbf{y}'(\xi) = I$, where I denotes the unit matrix. This implies that

$$\mathbf{y}'(\xi) = \left(\nabla^2 \Phi(\mathbf{y}(\xi))\right)^{-1}. \tag{3.18}$$

Define $\Psi(\xi) = \Phi(\mathbf{y}(\xi))$; it is also homogeneous function of degree 1. We have $\Psi(\xi) = \Phi(t, tx) = t\phi(x) = 0$ if ξ is the exterior normal to G at a point $\mathbf{y} \in \partial G$ close to $\mathbf{y}_0$; in particular, $\Psi(\xi_0) = 0$. The equation $\Psi(\xi) = 0$ defines a smooth cone G^* in $\mathbf{V}^*$, since $\nabla \Psi(\xi) = \nabla \Phi(\mathbf{y}(\xi))\, y'(\xi) \neq 0$ by virtue of (3.18).

Consider the embedding $e^* : \mathbf{E}^{n*} \to \left(\mathbf{E}^{n+1}\right)^*$ by $e^*(\xi) = (-1, \xi)$ and set $\Gamma^* = G^* \cap e^*(\mathbf{E}^{n*})$. The point ξ_0 belongs to $\Gamma^*$and the vector $\mathbf{y}_0$ is normal to G^* at ξ_0, since $\mathbf{y}_0 \nabla \Psi(\xi_0) = \mathbf{y}_0 \nabla \Phi(\mathbf{y}_0) = \Phi(\mathbf{y}_0) = 0$ and since Φ is homogeneous.

Therefore the vector x_0 is normal to Γ^* at ξ_0. Let K^* be the side of G^* in $\left(\mathbf{E}^{n+1}\right)^*$ for which $\mathbf{y}_0$ is the outward normal. The point $\eta = \xi_0 - \varepsilon\,(0,\xi)$ belongs to K^* for small $\varepsilon > 0$ and we have $\langle \eta, (0, x_0)\rangle = 1 - \varepsilon\,|x_0|^2 > 0$ for small ε. Further

$$\Psi(\eta) = \Phi(\mathbf{y}(\eta)) - \Phi(\mathbf{y}(\xi_0)) = -\varepsilon\,\langle \nabla\Phi(\mathbf{y}(\eta)), x_0\rangle + O\left(\varepsilon^2\right) = -\varepsilon + o\left(\varepsilon^2\right) < 0.$$

This implies that $\Psi < 0$ in K^* and $\Psi \geq 0$ in the complement to K^*. Differentiating the equation $\Psi(\xi) = \Phi(\mathbf{y}(\xi))$ yields

$$\nabla^2\Psi(\xi) = \mathbf{y}'(\xi)^t\, \nabla^2\Phi(\mathbf{y}(\xi))\, \mathbf{y}'(\xi) + \nabla\Phi(\mathbf{y}(\xi))\, \mathbf{y}''(\xi).$$

By (3.17) we find $\nabla\Phi(\mathbf{y}(\xi))\,\mathbf{y}''(\xi) = t\xi\,\mathbf{y}''(\xi) = 0$, since $\mathbf{y}'(\xi)$ is homogeneous in ξ of zero degree. Taking in account (3.18) we get

$$\nabla^2\Psi(\xi) = \left(\nabla^2\Phi(\mathbf{y}(\xi))\right)^* \tag{3.19}$$

where the symbol $*$ means the dual quadratic form. Denote by $\nabla^2_\Gamma\phi$ the restriction of the quadratic form $\nabla^2\Phi$ to the tangent plane of Γ:

$$T(\Gamma) \doteq \{\mathrm{d}x \in T(\mathbf{E}^n);\, \langle \nabla\phi, \mathrm{d}x\rangle = 0\}.$$

The pull back of this plane by the mapping $x = x(-1,\xi)$ coincides with the plane $\langle \nabla\Phi(1, x'(-1,\xi)), \mathrm{d}\xi\rangle = 0$. The last plane is just the plane $T(\Gamma^*)$ tangent to Γ^*, since $\nabla\Phi\,(1, x'(-1,\xi)) = \nabla\psi(\xi)$ where $\psi(\xi) \doteq \Psi(-1,\xi)$. Therefore (3.19) shows that the form $\nabla^2_{\Gamma^*}\psi(\xi)$ coincides with the dual to the form $\nabla^2_\Gamma\phi(x(\xi))$ whereas $\nabla\psi(\xi_0)$ is the exterior normal with respect to $D^* = K^*e^*(\mathbf{E}^*)$. The dual quadratic form has the same signature. This completes the proof. ►►

Definition. We call ψ the *Legendre dual* to the function ϕ.

Example 9. Let $\phi(a) = a(x) - c$ where $a(x) = 1/2a^{ij}x_ix_j$ is a non-singular quadratic form and c a constant. Calculate its Legendre dual function. We have $\Phi(t,x) = a(x) - ct^2$ and $\nabla\Phi = \left(-2ct, a^{ij}x_j\right)$. For any vector $\xi = (\xi_0, \xi),\ \xi \in \mathbb{R}^n$, we find $\mathbf{y}(\xi) = \left(-\xi_0/2ct, \sum_1^n b^{ij}\xi_j\right)$ from the equation $\nabla\Phi(\mathbf{y}) = \xi$, where $\left\{b^{ij}\right\}$ is the matrix inverse to $\left\{a^{ij}\right\}$. By definition, we have

$$\Psi(\xi) = \Phi(\mathbf{y}(\xi)) = \frac{1}{2}\sum_{i,j=1}^{n} b^{ij}\xi_i\xi_j - \frac{\xi_0^2}{4c}$$

and $\psi(\xi) = \Psi(-1,\xi) = b(x) - 1/4c$, where $b(\xi) = 1/2b^{ij}\xi_i\xi_j$ is the dual quadratic form.

Theorem 3.13 *Let $\Gamma = \partial D$ be a smooth hypersurface with the non-singular curvature form at a point x_0, f_λ be a function of the form (3.16) where a is supported by a small neighborhood U of x_0. We have, in a neighborhood of the dual hypersurface Γ^* to $\Gamma \cap U$,*

$$Mf_\lambda(\xi) = (2\pi)^{(n-1)/2}\,\frac{a(x(\xi)) + O(\psi(\xi))}{\left|\det \nabla^2_\Gamma\phi(x(\xi))\right|^{1/2}}\,g_\mu(\xi) + S,$$

where S is a smooth function and

(i) $$g_\mu(\xi) = (-1)^{q/2} \frac{\psi(\xi)_+^\mu}{\Gamma(\mu+1)}, \; \mu = \lambda + (n-1)/2,$$

if q is even, where ψ is the Legendre dual to ϕ;

(ii) $$g_\mu(\xi) = (-1)^{p/2} \frac{\psi(\xi)_-^\mu}{\Gamma(\mu+1)},$$

if p is even;

(iii) $$g_\mu(\xi) = (-1)^{(q+1)/2} \frac{\Gamma(-\mu)}{\pi} \left[\psi(\xi)_+^\mu + \cos(\mu\pi)\, \psi(\xi)_-^\mu\right],$$

if p and q are odd, and $\mu \neq 0, 1, 2, \ldots$, and

(iv) $$g_\mu(\xi) = \frac{(-1)^{\mu+(q-1)/2}}{\pi} \frac{\psi(\xi)^\mu}{\mu!} \ln|\psi(\xi)|,$$

if p, q are odd and $\mu \geq 0$ is an integer.

Remark. Let $\mu \to m$ for a natural m in the case (iii). Then $\Gamma(-\mu) \to \pm\infty$, whereas the difference

$$h \doteq g_\mu - (-1)^{\mu+(q-1)/2} \frac{\Gamma(-\mu)}{\pi} \psi(\xi)^m$$

has the finite limit as in (iv). The second term in the right side is smooth and can be included in the term S.

◀ Choose a point $x_0 \in \Gamma$ and set $\xi_0 = \nabla\phi(x_0)$; this is an exterior conormal vector to Γ at x_0. The normalized vector ξ_0 belongs to Γ^* and x_0 is an exterior normal to Γ^* at this point. The interior normal ray to Γ^* at ξ_0 is given by the equation $\xi = \xi_0 - tx_0, t \geq 0$ and $\psi = t + o(t)$. The value of the function Mf on this ray is equal to

$$Mf(t) = \int_{H(x_0,t)} a(x) \frac{\phi_-^\lambda(x)}{\Gamma(\lambda+1)} \frac{\mathrm{d}V}{\langle \xi_0, \mathrm{d}x\rangle}, \tag{3.20}$$

where $H(x_0, t)$, denotes the hyperplane $\{\langle x - x_0, \xi_0\rangle = t\}$ and $\mathrm{d}V$ is the Euclidean volume form in $\mathbf{E}$. We can take Euclidean coordinates $t = \langle x_0 - x, \xi_0\rangle, y_1, \ldots, y_{n-1}$ instead of x and replace ϕ by the function $\tilde{\phi}(t, y) \doteq \phi_0(y) - t$. By Morse's lemma we can change the coordinates y to z in such a way that $\phi_0(y) = Q(z) = 1/2 \sum \pm z_j^2$ is a quadratic form. Then we can write

$$Mf(t) = a(x_0) \int \frac{(Q(z) - t)_-^\lambda}{\Gamma(\lambda+1)} h(z)\, \mathrm{d}z + O(\psi),$$

where $h \in D(\mathbb{R}^n)$ is a function that equals 1 in a neighborhood of the origin; S is an integral of the form (3.20). Now we proceed by means of Fubini's theorem

$$\frac{1}{\Gamma(\lambda+1)} \int (Q(z) - t)_-^\lambda\, h(z)\, \mathrm{d}z = \frac{1}{\Gamma(\lambda+1)} \int \int_{Q=s} \frac{h(z)\, \mathrm{d}z}{\mathrm{d}Q} (t-s)_+^\lambda\, \mathrm{d}s$$

where the surface $\{Q = s\}$ is oriented by the form $\mathrm{d}Q$. Apply Proposition 3.14 to the inner integral. To complete the proof we need to evaluate the convolution

$$I(\lambda) = \int_0^t \frac{s^{(n-3)/2}}{\Gamma((n-1)/2)} \frac{(t-s)^\lambda}{\Gamma(\lambda+1)} \mathrm{d}s.$$

According to (1.7) we find

$$I(\lambda) = H^\lambda * H^{(n-1)/2} = H^\mu.$$

This implies the formulae (i) and (ii). The convolution

$$J = \int_{-\infty}^t \varepsilon(s) \frac{s^{(n-3)/2}}{\Gamma((n-1)/2)} \ln|s| \frac{(t-s)^\lambda}{\Gamma(\lambda+1)} \mathrm{d}s$$

is a bit more complicated and we have introduced the factor $\varepsilon \in S(\mathbb{R})$ which is equal to 1 in a neighborhood of the origin. This helps this integral to converge at infinity. We need to evaluate this integral up to a smooth term for the case when the number $k = (n-1)/2$ is natural. The convolution J can be written in the form

$$J = F^* \left(F\left(\frac{s_+^\lambda \mathrm{d}s}{\Gamma(\lambda+1)} \right) F\left(\frac{s^{k-1}}{\Gamma(k)} \varepsilon(s) \ln|s| \,\mathrm{d}s \right) \right).$$

By (1.5) and (1.2),

$$\begin{aligned} F\left(\frac{s_+^\lambda \mathrm{d}s}{\Gamma(\lambda+1)} \right) &= \mathrm{j}^{-\lambda-1} (\sigma - 0\imath)^{-\lambda-1} \\ &= \mathrm{j}^{-\lambda-1} \left(\sigma_+^{-\lambda-1} - \exp(-\lambda\pi\imath)\, \sigma_-^{-\lambda-1} \right), \\ F\left(\frac{s^{k-1}}{\Gamma(k)} \varepsilon(s) \ln|s| \,\mathrm{d}s \right) &= -\frac{1}{2} \mathrm{j}^{1-k} \left(\sigma_+^{-k} - \exp(k\pi\imath)\, \sigma_-^{-k} \right) * \tilde{\varepsilon}(\sigma). \end{aligned}$$

We can take ε such that $\tilde{\varepsilon}$ is compactly supported by $[-1,1]$. We can neglect the term $*\tilde{\varepsilon}$ for $|\sigma| > 1$. Therefore the product equals

$$-\frac{1}{2} \mathrm{j}^{-\lambda-k} \left(\sigma_+^{-\lambda-k-1} + \exp(-\pi(\lambda+k)\imath)\, \sigma_-^{-\lambda-k-1} \right)$$

for $|\sigma| > 1$. Apply the inverse Fourier transform and get, for $\mu = \lambda + k \neq 0, 1, 2, \ldots$

$$\begin{aligned} F\left(\sigma_+^{-\mu-1} \mathrm{d}\sigma \right) = \Gamma(-\mu)(-\mathrm{j})^\mu (s + 0\imath)^\mu, \; F\left(\sigma_-^{-\mu-1} \mathrm{d}\sigma \right) \\ = \Gamma(-\mu)\, \mathrm{j}^\mu (s - 0\imath)^\mu, \end{aligned}$$

$$\begin{aligned} 2J &= -\mathrm{j}^{-\mu} F^* \left(\sigma_+^{-\mu-1} + \exp(-\mu\pi\imath)\, \sigma_-^{-\mu-1} \right) \\ &= \Gamma(-\mu) \exp(-\pi\mu\imath) \left[(s+0\imath)^\mu + (s-0\imath)^\mu \right] \\ &= -2\Gamma(-\mu) \exp(-\pi\mu\imath) \left[s_+^\mu + \cos(\mu\pi)\, s_-^\mu \right]. \end{aligned}$$

This implies (iii).

To check (iv) for $\mu = m$ we add $J(\mu) = \Gamma(-\mu)\exp(-\pi\mu\imath)\, s^m$ and take $\mu \to m$ for a natural m. We get

$$\begin{aligned} I_2 + J &= -\Gamma(-\mu)\exp(-\pi\mu\imath)\left[s_+^\mu + \cos(\mu\pi)\, s_-^\mu - s^m\right] \\ &= -\Gamma(-\mu)(\mu - m)\exp(-\pi\mu\imath)\left[\frac{s_+^\mu - s_+^m}{\mu - m} + \frac{\cos(\mu\pi)\, s_-^\mu - \cos(m\pi)\, s_-^m}{\mu - m}\right] \\ &\to \frac{s^m}{m!}\ln|s|, \end{aligned}$$

since $\Gamma(-\mu)(\mu - m) \to \mathrm{res}_{-m}\Gamma = (-1)^{m-1}/m!$. ▶

3.8 Nonlinear artifacts

Suppose we have corrupted projection data $T(Rf)$ instead of Rf where T is a nonlinear function in $\mathbb{R}$. Suppose for simplicity that $T = T(s)$ is linear in the interval $(-\infty, S)$, smooth and bounded for $s \geq S$ (we call T the truncation function). If we substitute the corrupted data in the reconstruction operator of Chapter 2, we get a distorted image $\tilde{f} = R^\sharp KT(Rf)$, $K = (-\partial_p \mathbf{H}/2\pi)^{n-1}$, instead of f. Some specific artificial details can appear in $\tilde{f}$, which we call nonlinear artifacts. Some of the nonlinear artifacts have clear geometrical structure, at least, for model functions f as in the previous section. Suppose for simplicity that $n = 2$ and take the function f in the form (3.16) with $\lambda = 0$, that is $f = g\delta(\phi)$ where $g > 0$ is a smooth function. According to Theorem 3.13, we have locally $Rf = h\psi^{-1/2} + h_0$ for smooth functions h and h_0 where $h > 0$ since $q = 0$. The truncation $T(Rf)$ does not affect the term h_0 if the threshold S is big enough and changes the profile of the singular function $\psi^{-1/2}$ "uniformly" over Γ^*. So the result will have the form $T(Rf) \approx \tilde{\psi} + h_0$ where $\tilde{\psi}$ has a more or less large gradient close to Γ^*. If the threshold of truncation is big enough, we can, nevertheless, recognize the shape of Γ from the local corrupted reconstruction $\tilde{f}$. More complicated artifacts are caused by the global structure of Γ that generates singularities in the dual curve Γ^*. The action of truncation may be more strong in neighborhoods of these points. The following pictures show some simple situations.

Fig. 3.2 shows the cardioid-like curve Γ and the dual curve Γ^* shown in thick type. (The dual curve is obtained by means of the spherical duality and the central projection from sphere to the plane as in Section 3.2.) According to Theorem 3.13 the function Rf increases faster near two cusp points of Γ^*, where the curvature of the curve tends to infinity. These provoke strong line artifacts at the tangent lines to Γ at the inflection points. One more artifact line is the double tangent to Γ which is caused by amplification of Rf at the self-intersection point of Γ^*, see Fig. 3.3.

Fig. 3.4 shows the curve Γ consisting of two circles. The curve shown in thick type is the dual curve with respect to the origin at the bottom of the picture. Four self-intersection points of Γ^* correspond to four common tangents shown by dotted lines. These are locations of nonlinear artifacts.

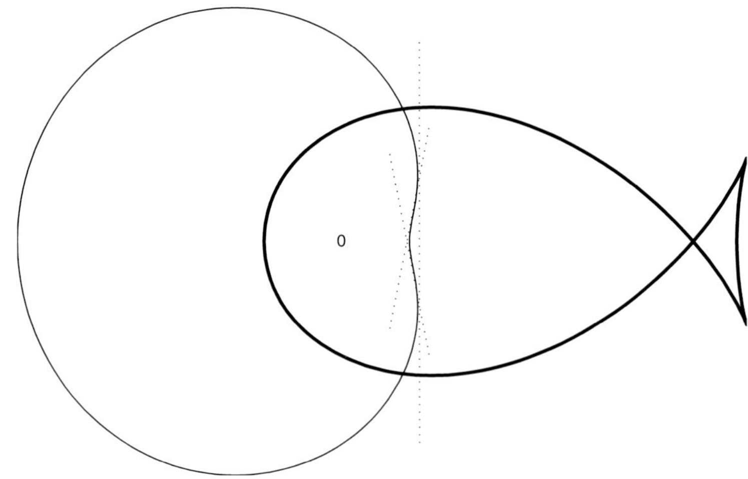

Figure 3.3: Artifacts from two inflection tangents and a double tangent line

The truncation function T is model of a nonlinear measurement process in a standard CT technique: each measurement can be modelled by the function $\exp\left(-\int f\mathrm{d}S\right)$ where the integral is taken over a thin strip from the source to the detector. Therefore if the integral is big, the measurement is small and hardly distinguishable because of the noise, which means that big values of Rf are, in fact, truncated. The algorithms of type "filtered backprojection" (used as a standard software) have the same structure $R^{\sharp}KR$ with a different "regularized" kernel K whereas the projection data Rf is discrete and noised. Nevertheless the artifacts that sometimes appear in a tomogram, have strong similarity to that we are discussing for the continuous model. See more details in [67].

3.9 Pizetti formula for arbitrary signature

Let $Q = 1/2\sum a^{ij}x_i x_j$ be a non-singular quadratic form in $X = \mathbb{R}^n$ of signature (p,q). Define the dual differential operator $Q^*(D) = 1/2\sum a_{ij}\partial_i\partial_j$ where $\{a_{ij}\} = \left\{a^{ij}\right\}^{-1}$. Choose an orientation of X and consider the family of hypersurfaces $X(s) = \{Q(x) = s\},\ s \in \mathbb{R}$ with the orientation defined by $\mathrm{d}Q$. Introduce the sequence of generalized functions in X supported at the origin

$$F_k(\omega) \doteq \frac{(2\pi)^{n/2}}{|\Delta|^{1/2}} \frac{Q^*(D)^k \phi(0)}{\Gamma\left(\frac{n}{2}+k\right)},\ k = 0,1,2,\ldots;\ \omega = \phi\mathrm{d}x.$$

Proposition 3.14 *For an arbitrary smooth n-form ω with compact support in $\mathbb{R}^n$, we have the following asymptotic expansions:*

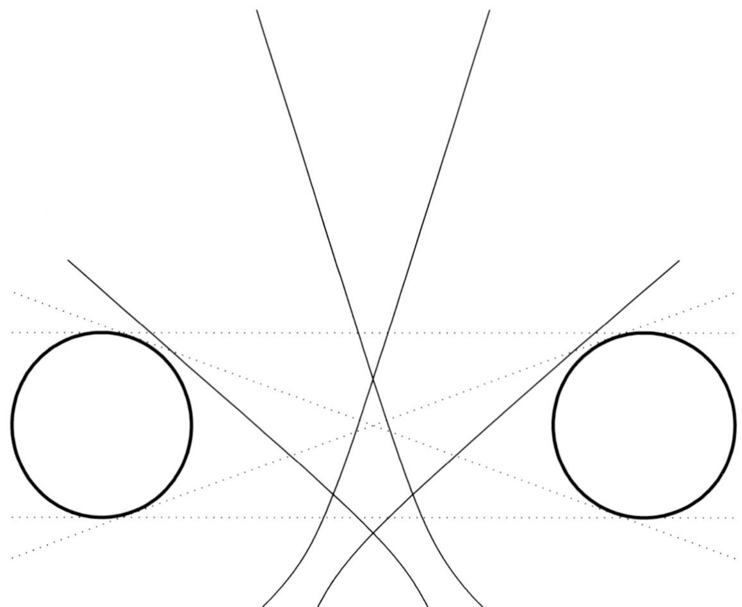

Figure 3.4: Artifacts of two disjoint circles

(i) *for $p+q$ odd,*

$$\int_{X(s)} \frac{\omega}{\mathrm{d}Q} \sim \sum_{k\geq 0} b_k^{\pm} |s|^{n/2-1} \frac{s^k}{k!} + r(s),$$

where $b_k^+ - b_k^- = (-1)^{q/2} F_k(\chi)$, if q even and $b_k^+ - b_k^- = (-1)^{p/2} F_k(\chi)$, if p even;

(ii) *for p and q even,*

$$\int_{X(s)} \frac{\omega}{\mathrm{d}Q} \sim \sum_{k\geq 0} b_k^{\pm} s^{n/2-1} \frac{s^k}{k!} + r(s),$$

where $b_k^+ - b_k^- = (-1)^{q/2} F_k(\chi)$;

(iii) *for p, q odd,*

$$\int_{X(s)} \frac{\omega}{\mathrm{d}Q} \sim \sum_{k\geq 0} b_k s^{n/2-1} \ln|s| \frac{s^k}{k!} + r(s),$$

where $b_k = (-1)^{(q-1)/2} \pi^{-1} F_k(\chi)$ and r is a smooth function which vanishes at $s=0$ together with all derivatives. These expansions can be derived term by term.

◀ For a proof we apply the stationary phase method to the integral with the quadratic phase function:

$$\int_X \exp(\mathrm{j}tQ(x))\,\phi\mathrm{d}x = \frac{e_{p,q}(t)}{|t\,\Delta|^{1/2}}\sum_{j=0}^{N}\frac{1}{j!}\left(-\frac{Q^*(D)}{\mathrm{j}t}\right)^j\phi(0) + \frac{R_N(\omega,t)}{t^{N+(n+1)/2}}, \quad (3.21)$$

where $e_{p,q}(t) \doteq \exp(\operatorname{sgn} t\,\pi\imath\,(p-q)/4)$ and the remainder R_N can be uniformly estimated in terms of the $2N+1$th derivatives of ϕ. By Fubini's theorem we can write

$$\int_X \exp(\mathrm{j}tQ(x))\,\phi\mathrm{d}x = \int_{-\infty}^{\infty}\exp(\mathrm{j}st)\,\mathrm{d}s\int_{X(s)}\frac{\omega}{\mathrm{d}Q}$$

and we can recover the Funk transform of the Leray form by applying the Fourier transform to (3.21). First we multiply (3.21) by $1-h$ where $h \in D(\mathbb{R})$ is an arbitrary function that equals 1 near the origin; the singularity of (3.21) at the origin is not relevant to the singularity of the Funk transform at $s=0$. Applying the Fourier transform yields:

$$\int_{X(s)}\frac{\omega}{\mathrm{d}Q} = \sum_{i=0}^{N}\frac{1}{i!}(Q^*(D))^i\,\phi(0)$$
$$\times\int_{-\infty}^{\infty}\exp(-\mathrm{j}ts)\,\frac{(1-h(t))\,e_{p,q}(t)}{|t\,\Delta|^{1/2}}(-\mathrm{j}t)^{-i}\,\mathrm{d}t + \tilde{R}(s),$$

where $\tilde{R} \in C^{N+n/2-1}(\mathbb{R})$. Calculating the Fourier integrals by means of Examples 10, 11 of Section 1.4, we complete the proof. ▶

Chapter 4

Reconstruction from Line Integrals

4.1 Pencils of lines and John's equation

Let $\mathbf{E}$ be a Euclidean space of dimension n and f be a function in $\mathbf{E}$ that decreases at infinity in such a way that any line integral of f converges absolutely. It is sufficient to assume that $f(x) = O\left(|x|^{-1-\varepsilon}\right)$ at infinity for some $\varepsilon > 0$. We denote by g the line integral of f :

$$g(x, \theta) = \int_{L(x,\theta)} f \, \mathrm{d}s, \tag{4.1}$$

where $L(x, \theta)$ denotes the straight line (or the ray) through the point $x \in \mathbf{E}$ that is parallel to the unit vector θ. In the case $n = 2$, data of the line integrals defines the Radon transform of f as follows: $Rf(p, \omega) = g(x, \theta)$ where θ is orthogonal to the unit vector ω and x is any point such that $\langle \omega, x \rangle = p$. In the case $n > 2$, the manifold $\mathsf{A}_1(\mathbf{E})$ of all straight lines has dimension $2n - 2$ which implies that data of all line integrals are *redundant*. To avoid redundancy we state the reconstruction problem as follows:

Problem: to find a reconstruction formula $g|\Sigma \mapsto f$ for functions f supported by a compact set $K \subset \mathbf{E}$ for a submanifold $\Sigma \subset \mathsf{A}_1(\mathbf{E})$ of lines (rays) of dimension n. We call such a manifold a *pencil*. The data of $g(L)$ for a pencil of lines have no dimension redundancy. The pencil Σ should, of course, fulfil the completeness condition for K (see Chapter 6).

THE CASE $n = 3$ is important for applications to X-ray tomography. There are several classes where explicit reconstruction formulae are known:

1. Choose a plane $H \subset \mathbf{E}$ and consider the pencil Σ_H of straight lines that are parallel to H. Take an arbitrary plane H' that is parallel to H. Any line $L \subset H'$ belongs to Σ_H, hence we know the line transform $g(L)$. Apply the inversion of the Radon transform in H' and reconstruct the function $f : H' \to \mathbb{C}$ for each plane H' that is parallel to H.

2. Take a curve $\mathbf{C} \subset \mathbb{P}(\mathbf{E})$ such that any plane $H \subset \mathbf{E}$ has non-empty intersection with $\mathbf{C}$ at infinity. Consider the pencil $\Sigma(\mathbf{C})$ of lines L that meet $\mathbf{C}$ at infinity. There exists a reconstruction method for this pencil. Indeed, take a plane H; let $c \in \mathbf{C}$ be a point where H meets $\mathbf{C}$. Any line $L \subset H$ that contains the point c at infinity belongs $\Sigma(\mathbf{C})$. Such lines L are parallel one to another and makes a foliation of H. Let $T \subset H$ be a line that is orthogonal to L. By Fubini's theorem

$$\int_H f \mathrm{d}S = \int_T \mathrm{d}t \int_L f \mathrm{d}s,$$

where $\mathrm{d}t$ is the Euclidean density in T. Thus we know the Radon transform $Rf\,(H)$ for any 2-plane H in $\mathbf{E}$ and can reconstruct the function f. This reconstruction formula is given by four-fold integration. The method of the next section uses two-fold integration.

3. Let Γ be a curve in $\mathbf{E}$ and $\Sigma(\Gamma)$ be the pencil of rays with vertices in Γ. A function f with compact support can be reconstructed if the completeness condition is fulfilled. The pencil $\Sigma\,(\Gamma)$ is characteristic. A reconstruction can be done by two-fold integration.

Note that the pencil Σ_H as in Class **1** is equal to the pencil $\Sigma\,(C)$ where $C = H \cap \mathbb{P}\,(\mathbf{E})$ is an improper line. Class **2** can be reduced to Class **3**, since the curve C can be transported to $\mathbf{E}$ by a suitable projective transformation P. The mapping P transforms lines to lines, hence, $P\,(\Sigma\,(C)) = \Sigma\,(P\,(C))$. By Proposition 3.2 a reconstruction formula for $\Sigma\,(C)$ can be translated to a reconstruction formula for $\Sigma\,(P\,(C))$ and vice versa.

4. Let S be a surface in $\mathbf{E}$ and $\Sigma(S)$ be the pencil of rays with vertices in S that are tangent to S. It is also characteristic. A reconstruction of a function f with compact support is possible under the completeness condition. Class **3** is, in a sense, contained in the closure of Class **4**. Indeed, take the ε-neighborhood Γ_ε of a smooth curve Γ. The surface $\partial\Gamma_\varepsilon$ is smooth if ε is small enough. The pencil $\Sigma\,(\partial\Gamma_\varepsilon)$ tends to $\Sigma\,(\Gamma)$ as $\varepsilon \to 0$ and the reconstruction formula of Class **4** has a limit which gives a reconstruction for the pencil $\Sigma\,(\Gamma)$.

John's equations. The manifold of lines $\mathsf{A}_1\,(\mathbf{E}^n)$ has dimension $2n-2$, hence $g\,(L)$ is a function of $2n-2$ variables in any local chart of $\mathsf{A}_1\,(\mathbf{E})$ whereas f is a function of n variables. The function g satisfies a system of differential equations which express redundancy of variables. To write these equations we choose the coordinates $x_1, \ldots, x_n$ in $\mathbf{E}$ and take the chart F in $\mathsf{A}_1\,(\mathbf{E})$ that contains all straight lines L that are not parallel to the plane $x_n = 0$ in $\mathbf{E}$. Write $x = (x', x_n)$ where $x' = (x_1, \ldots, x_{n-1})$. Take the point $(x, 0) \in L$ and the vector $(\theta, 1)$ parallel to L; the vectors $x, \theta \in \mathbb{R}^{n-1}$ are coordinates of L in the chart F. These coordinates parameterize the line integral

$$g(x, \theta) = \int_{L(x,\theta)} f \mathrm{d}s.$$

Proposition 4.1 *The function g satisfies John's equations*

$$\frac{\partial^2 h}{\partial x_i \partial \theta_j} = \frac{\partial^2 h}{\partial x_j \partial \theta_i}, \; i, j = 1, \dots, n-1, \tag{4.2}$$

where $h(x, \theta) \doteq \|(\theta, 1)\|^{-1} g(x, \theta)$.

◄ The line L is given by parametric equations $x' = x + t\theta,\ x_n = t$; the Euclidean line density in L is equal to $\mathrm{d}s = \|(\theta, 1)\| \, \mathrm{d}t$, consequently

$$h(x, \theta) = \int f(x + t\theta, t) \mathrm{d}t.$$

Therefore

$$\frac{\partial^2 h}{\partial x_i \partial \theta_j} = \int_{-\infty}^{\infty} \frac{\partial^2 f}{\partial y_i \partial y_j} (x + t\theta, t) \, t \mathrm{d}t.$$

We get the same integral formula for the function $\partial^2 h / \partial x_j \partial \theta_i$. ►

John's equations (4.2) also hold for the line transform to the projective space $\mathbb{P}^n$. For this we consider f as a homogeneous function on $\mathbf{E}^{n+1} \setminus \{0\}$ of degree -2. The line transform $g(L)$ of f is well defined for all projective lines L in $\mathbb{P}^3$ (see Remark in Section 5.5). The function g still fulfils John's equation in any chart of the manifold of lines which is the Grassmann manifold $\mathbf{Gr}_2 \left(\mathbf{E}^{n+1} \right)$. John's equations belong to neither elliptic nor hyperbolic type.

The reconstruction problem can be thought as a boundary value problem for solutions g of John's equation where the "boundary" is our pencil Σ and only the value of g on Σ is known. If Σ is the boundary of a domain Δ in A_1, this is the Dirichlet problem which is, in general, hard to study since the equation is non-elliptic. On the other hand, there is an obstacle for local solution of the second order equation, since no transversal derivative of g is known on Σ. The only chance to solve such a boundary problem is to take a pencil Σ that is characteristic with respect to John's operator $J = \partial^2 / \partial x_i \partial \theta_j - \partial^2 / \partial x_j \partial \theta_i$ at each of its points. This is in fact, the case for all the Classes **1**–**4**. Recall that a hypersurface given by the equation $\phi(x) = 0$ in a space $\mathbf{V}$ is called *characteristic* with respect to a linear differential operator $a(x, D)$, if $\sigma(x, \nabla \phi(x)) = 0$ where $\sigma(x, \xi)$ is the principal symbol of a, i.e., the sum of terms of a of maximal degree m where each derivative D^j is replaced by $(\imath \xi)^j$.

Proposition 4.2 *Any pencil of the Class* **1**, **2**, **3** *or* **4** *is characteristic with respect to any John's equation.*

◄ Classes **1** and **2** are subclasses of **3** defined for the manifold of lines in $\mathbb{P}^3$. It is sufficient to check the statement for the Class **4**, it holds for Class **3** by continuity. Take a surface $S \subset \mathbf{E}^3$ and a point $p \in S$. Let L be a tangent line to S at p such that the normal curvature of S does not vanish in the direction of L. Then the pencil $\Sigma(S)$ is a smooth submanifold of $\mathsf{A}_1(\mathbf{E})$ in a neighborhood of L.

Take a pair of parallel planes P, Q and a coordinate system (x_1, x_2, x_3) as above to define a local chart in $\mathsf{A}_1(\mathbf{E})$. These planes are not parallel to L and we can take P through the point p. There tangent space $T_L(\Sigma(S))$ is of dimension 3 and there is freedom for the vector (θ_1, θ_2). On the other hand, if we move the point $p \in S$ in the direction θ, the variation of coordinates $x_1(L), x_2(L)$ will be of order $o(|t|)$. If we move p by an arbitrary vector t tangent to S, the point $L \cap P$ moves in a direction orthogonal to $(\phi_1', \phi_2', 0)$. It follows that the covector $(\phi_1', \phi_2', 0, 0)$ is orthogonal to the tangent space $T_L(\Sigma(S))$. We now need only to check that the principal symbol σ of John's equation written in this chart vanishes on this covector. We have $\sigma(\eta_1, \eta_2, \omega_1, \omega_2) = -\eta_1\omega_2 + \eta_2\omega_1$, which makes the statement obvious. ▶

Remark. By [39] the inverse statement is true: any generic 3-submanifold in $\mathsf{A}_1(\mathbf{E})$ which is characteristic with respect to John's operator coincides locally with a pencil of Class **4**. Proposition 4.2 makes plausible that an explicit solution of the reconstruction problem can be found for Classes **3** and **4**. It gives, however, no tool to find such explicit formulae. In Sections 7.7–7.9 we meet a similar situation, where Darboux's equation will be the master one. There we shall use some specific tools of theory of partial differential equations.

4.2 Sources at infinity

Let $\mathbf{E} = \mathbf{E}^n$ be a Euclidean space of arbitrary dimension $n > 1$ and $\mathbb{P}(\mathbf{E})$ be the projective $n-1$-space of improper points of $\mathbf{E}$, i.e., space of lines through the origin. Let $\mathbf{S}$ be the unit sphere in $\mathbf{E}$; we have the natural surjection $\mathbf{p} : \mathbf{S} \to \mathbb{P}(\mathbf{E})$, which glues together each pair of opposite points of the sphere. Take a curve $\mathbf{C}$ in $\mathbb{P}(\mathbf{E})$ and consider the pencil $\Sigma = \Sigma(\mathbf{C})$ of lines in $\mathbf{E}$ that meets at infinity $\mathbf{C}$. We reconstruct a function f in $\mathbf{E}$ with compact support from line integrals (4.1) for lines $L \in \Sigma(\mathbf{C})$.

Theorem 4.3 *Let $\mathbf{C} \subset \mathbb{P}(\mathbf{E})$ be a closed curve of the class C^1 that is not homotopic to a point, $\theta = \theta(s), 0 \le s \le S$ be a parametrization of $\mathbf{C}$ such that $|\theta'| = 1$. The formula*

$$f(x) = -\frac{1}{2\pi^2} \int_0^S \mathrm{d}s \int_{-\infty}^{\infty} \frac{\partial}{\partial q} g(x + q\theta'(s), \theta(s)) \frac{\mathrm{d}q}{q} \tag{4.3}$$

gives a reconstruction for a function $f \in C^2(\mathbf{E})$ of compact support from the data of line integrals $g(L)$, $L \in \Sigma(\mathbf{C})$. The principal value of the interior integral is meant.

◀ Write the Fourier transform of an unknown function

$$\hat{f}(\xi) = \int_{\mathbf{E}} \exp(-\mathrm{j}\langle \xi, x\rangle) f(x) \mathrm{d}x,$$

where $\mathrm{d}x$ is the Euclidean volume density. Take a unit vector θ that is orthogonal to ξ and write $x = y + t\theta$, where $\langle y, \theta \rangle = 0$. Integrate consecutively over t and $y \in \theta^{\perp}$, taking into account that $\langle \xi, x \rangle = \langle \xi, y \rangle$:

$$\hat{f}(\xi) = \int_{\theta^{\perp}} \exp(-\mathrm{j} \langle \xi, y \rangle) g(y, \theta) \mathrm{d}y, \tag{4.4}$$

where $\mathrm{d}y$ is the Euclidean area element in the hyperplane $\theta^{\perp}$.

First note that the fundamental group $\pi_1(\mathbb{P}(\mathbf{E}))$ is isomorphic to $\mathbb{Z}_2$ and any projective line $\mathbf{L}$ is a generator of this group. The topological condition means that the curve $\mathbf{C}$ is homotopicaly equivalent to $\mathbf{L}$. Fix a point $c_0 \in \mathbf{C}_*$ and choose the smooth curve $\mathbf{C}_+ \subset \mathbf{S}$ that starts with the point c_0 such that the mapping $\mathbf{p} : \mathbf{C}_+ \to \mathbf{C}$ is a bijection. The end of $\mathbf{C}_+$ is the point $-c_0$, otherwise the curve $\mathbf{C}_+$ would be closed in $\mathbf{S}$ and homotopic to a point. The mapping π transforms any homotopy $\mathbf{C}_+ \sim \mathrm{pt}$ into a homotopy $\mathbf{C} \sim \mathrm{pt}$, which does not exist due to the condition. Thus we have $\mathbf{C}_* = \mathbf{C}_+ \cup \mathbf{C}_-$ where $\mathbf{C}_- \doteq -\mathbf{C}_+$. Take the parameterization $\theta = \theta(s)$ of the curve $\mathbf{C}_+$ such that $\theta(0) = c_0, \theta(S) = -c_0$. The point $\theta(s) = -\theta(s - S)$ runs for $s \in [S, 2S]$ over the curve $\mathbf{C}_-$. Then $\theta = \theta(s)$, $s \in [0, 2S]$ is a C^1-parameterization of the curve $\mathbf{C}_*$ such that $\theta(2S) = \theta(0)$. We have $\langle \theta'(s), \theta(s) \rangle = 0$ since θ and θ' are unit vectors. Choose a Euclidean coordinate system $\xi = (\xi_1, \ldots, \xi_n)$ in the dual space $\mathbf{E}^*$ and orientate $\mathbf{E}^*$ by means of the volume form $\upsilon \doteq \mathrm{d}\xi_1 \wedge \cdots \wedge \mathrm{d}\xi_n$.

Consider the $n+1$-manifold $\mathbf{C}_* \times \mathbf{E}^*$ oriented by the form $\upsilon \wedge \mathrm{d}s$. The restriction of the form $\mathrm{d} \langle \theta(s), \xi \rangle$ to $\mathbf{C}_* \times \mathbf{E}^*$ is equal to $\langle \theta(s), \mathrm{d}\xi \rangle + \langle \theta'(s), \xi \rangle \mathrm{d}s$ and does not vanish. Take a differential form ω on $\mathbf{C}_* \times \mathbf{E}^*$ that fulfils the equation

$$\omega \wedge \mathrm{d} \langle \theta(s), \xi \rangle = \upsilon \wedge \mathrm{d}s. \tag{4.5}$$

Such a form is defined uniquely up to a term $\chi \wedge \mathrm{d} \langle \theta(s), \xi \rangle$ for an $n-1$-form χ. Therefore the restriction of ω to the submanifold

$$\Gamma \doteq \{(s, \xi) : \ \langle \theta(s), \xi \rangle = 0, \ \langle \theta'(s), \xi \rangle \geq 0\}$$

is uniquely defined since the form $\mathrm{d} \langle \theta(s), \xi \rangle$ vanishes in Γ. We define an orientation on Γ by means of the form ω and consider the mapping $\gamma_{\mathbf{C}} : \Gamma \to \mathbf{E}^*$, $\gamma_{\mathbf{C}}(s, \xi) = \xi$. The pair $\Pi = (\Gamma, \gamma_{\mathbf{C}})$ is an odd n-chain in $\mathbf{E}^*$.

Lemma 4.4 *The chain* Π *is closed.*

◀ For any s the half-space $\Gamma(s) \doteq \{\xi : \langle \theta(s), \xi \rangle = 0, \langle \theta'(s), \xi \rangle \geq 0\}$ is complementary to $\Gamma(s+S) = \{\langle \theta'(s+S), \xi \rangle \geq 0\}$ and their orientations are given by the same form ω. This implies that the boundaries of these half-planes have opposite orientation and the lemma follows. ▶

The mapping $\gamma_{\mathbf{C}}$ is proper and its topological degree $\deg \gamma_{\mathbf{C}}$ is well defined.

Lemma 4.5 $\deg \gamma_{\mathbf{C}} = 1$.

◀ Denote this mapping by $\gamma_{\mathbf{C}}$. The degree of $\gamma_{\mathbf{C}}$ is invariant under any deformation of the curve $\mathbf{C}$. We have a homotopy $\mathbf{C} \sim \mathbf{L}$ in this class where $\mathbf{L}$ is an arbitrary projective line in $\mathbb{P}(\mathbf{E})$. Therefore we have $\deg \gamma_{\mathbf{C}} = \deg \gamma_{\mathbf{L}}$. The mapping $\gamma_{\mathbf{L}}$ restricted to the interior of Γ is a bijection onto $\mathbf{E}^* \backslash \mathbf{L}^{\odot}$ where $\mathbf{L}^{\odot}$ is the orthogonal complement to the plane that contains $\mathbf{L}$. The form $\omega' = \upsilon / \langle \theta', \xi \rangle$ fulfils the equation (4.5), which implies that ω defines the orientation of Γ. It coincides with the orientation of $\mathbf{E}$ since $\langle \theta', \xi \rangle > 0$ in Γ. This yields the equation $\deg \gamma_{\mathbf{L}} = 1$. ▶

Lemma 4.5 implies that

$$f(x) = \int_{\mathbf{E}} \exp(\mathrm{j} \langle \xi, x \rangle) \hat{f}(\xi)\, \upsilon = \int_{\Gamma} \exp(\mathrm{j} \langle \xi, x \rangle) \hat{f}(\xi) \gamma^* (\upsilon), \tag{4.6}$$

where $\gamma^* (\upsilon)$ is the pull back of υ under the mapping $\gamma = \gamma_{\mathbf{C}}$. Take an $n-1$-form ψ in $\mathbf{C}_* \times \mathbf{E}^*$ that satisfies $\psi \wedge \langle \theta, \mathrm{d}\xi \rangle = \upsilon$ and set $\omega = -\langle \theta', \xi \rangle\, \psi \wedge \mathrm{d}s$. We claim that $\gamma^* (\upsilon) = \omega$ in Γ. Indeed, the form $\rho \doteq \upsilon - \omega$ defined in $\mathbf{C}_* \times \mathbf{E}^*$ fulfils the equation

$$\begin{aligned} \rho \wedge \mathrm{d} \langle \theta(s), \xi \rangle &= \langle \theta', \xi \rangle\, (\upsilon \wedge \mathrm{d}s + \psi \wedge \mathrm{d}s \wedge \langle \theta, \mathrm{d}\xi \rangle) \\ &= \langle \theta', \xi \rangle\, (\upsilon - \psi \wedge \langle \theta, \mathrm{d}\xi \rangle) \wedge \mathrm{d}s = 0. \end{aligned}$$

This implies that ρ is a multiple of the form $\mathrm{d} \langle \theta(s), \xi \rangle$ which vanishes in Γ. We replace $\gamma^* (\upsilon)$ by ω in (4.6) and substitute (4.4):

$$\begin{aligned} &\int_{\Gamma} \exp(\mathrm{j} \langle \xi, x \rangle)\, \hat{f}(\xi)\, \gamma^* (\upsilon) \\ &= -\int_0^{2S} \mathrm{d}s \int_{\Gamma(s)} \langle \theta'(s), \xi \rangle \exp(\mathrm{j} \langle \xi, x - y \rangle)\, \psi \int_{\theta(s)^{\perp}} g(y, \theta(s)) \mathrm{d}y. \end{aligned} \tag{4.7}$$

For any fixed $s \in [0, 2S]$ the form $-\langle \theta', \xi \rangle\, \psi$ defines the orientation of the $n-1$-plane $\Gamma(s)$. We can write $\psi = \mathrm{d}\sigma \wedge \tau$, where $\sigma = \langle \theta'(s), \xi \rangle$, and τ is the volume form in the $n-2$-space $\mathbf{E}^*(s) = (\theta(s), \theta'(s))^{\perp}$. We have $y - x = p\theta + q\theta' + r$, where $p = \langle y - x, \theta \rangle$, $q = \langle y - x, \theta' \rangle$, $r \in \mathbf{E}(s)$, which yields $\langle \xi, x - y \rangle = -\sigma q + \langle \eta, r \rangle$, where η is the orthogonal projection of ξ to $\mathbf{E}^*(s)$. Substituting in the inner integral yields

$$\begin{aligned} \int_{\Gamma(s)} \langle \theta', \xi \rangle \exp(\mathrm{j} \langle \xi, x - y \rangle) \psi &= -\int_0^{\infty} \sigma \exp(-\mathrm{j}\sigma q)\, \mathrm{d}\sigma \int_{-\infty}^{\infty} \exp(\mathrm{j} \langle \eta, r \rangle)\, \tau \\ &= \frac{1}{(2\pi q - 0\imath)^2} \delta(r). \end{aligned}$$

We have $\delta(r)\, \mathrm{d}y = \mathrm{d}q$ in the right side of (4.7) and $y = x + p\theta + q\theta'$ since $r = 0$. Substituting in (4.7) gives the quantity

$$-\int_0^{2S} \mathrm{d}s \int_{\mathbb{R}} \frac{1}{(2\pi q - 0\imath)^2} g(y, \theta) \mathrm{d}q = -\frac{1}{4\pi^2} \int_0^{2S} \mathrm{d}s \int_{\mathbb{R}} \frac{1}{(q - 0\imath)^2} g(x + q\theta', \theta) \mathrm{d}q.$$

We removed the term $p\theta$ in the argument of g, which does not change this function. Now integrate by parts and apply (4.6):

$$f(x) = -\frac{1}{4\pi^2}\int_0^{2S} \mathrm{d}s \int_{\mathbb{R}} \frac{1}{q - 0\imath}\frac{\partial}{\partial q} g(x + q\theta', \theta)\mathrm{d}q.$$

Sum up the contributions of points s and $s + S$ for $s \in [0, S]$ using the relations

$$\theta(s + S) = -\theta(s),\ \theta'(s + S) = -\theta'(s),$$

$$\frac{\partial}{\partial q} g(x - q\theta', -\theta) = -\frac{\partial}{\partial q} g(x + q\theta', \theta),\ \frac{1}{q + 0\imath} + \frac{1}{q - 0\imath} = 2\left[\frac{1}{q}\right].$$

This yields (4.3). ▶

Remark 1. The interior integral (4.3) sums up the ray data measured from lines $L \subset P(x, s)$ where $P(x, s)$ is the plane through a point x that is parallel to a tangent plane to the curve $\mathbf{C}_* \subset \mathbf{S}$ at the point $\theta(s)$. The condition of compactness of supp f can be weakened. Really let us consider the following continuous mapping $\sigma : \mathbf{E} \times \mathbf{C} \to \Sigma(\mathbf{C})$ that sends a pair (x, c) to the line λ that passes through x and c. We say the a set $B \subset \mathbf{E}$ is $\mathbf{C}$-compact if the restriction of σ on $B \times \mathbf{C}$ is a proper mapping. This condition implies that the intersection of B with any tangent plane P is a compact set. The ray transform $g\,|\Sigma(\mathbf{C})$ is well defined for any continuous function f such that the set supp f is $\mathbf{C}$-compact. It is easy to see that the proof of Theorem 4.3 is still valid.

Remark 2. If $\mathbf{C}$ is a projective line, $\Sigma(\mathbf{C})$ is the manifold of lines that are parallel to the plane $H \subset \mathbf{E}$ such that $\mathbb{P}(H) = \mathbf{C}$. Then $g\,|\Sigma\,(\mathbf{C})$ is the two-dimensional Radon transform applied to each plane parallel to H.

4.3 Reduction to the Radon transform

Let f be a bounded function with compact support in $\mathbf{E}^3$; the family of ray integrals

$$g\,(y, \theta) = \int_0^{\infty} f(y + t\theta))\mathrm{d}t \tag{4.8}$$

defined for points $y \in \mathbf{E}$ and unit vectors $\theta \in \mathbf{E}$ is called an X-ray transform of f. The first derivative of Radon transform Rf can be evaluated in terms of the ray transform by Grangeat's method:

Theorem 4.6 *Let H be a plane in $\mathbf{E}^3$, $y \in H$ and $f \in C^2(\mathbf{E})$ be arbitrary function such the set* supp $f \cap H$ *is compact. Then we have*

$$\frac{\partial}{\partial p} Rf(H) = \int_0^{2\pi} \frac{\partial}{\partial q} g\,(y, q\omega + rv\,(\phi))\,|_{q=0}\mathrm{d}\phi, \tag{4.9}$$

where $v = v\,(\phi)$ runs along the unit circle in the plane $H - y$, and $r = \left(1 - q^2\right)^{1/2}$.

◀ We use the standard parameterization of planes $H = H(p,\omega) = \{x; \langle \omega, x\rangle = p\}$ in $\mathbf{E}$; we have $H - y = H(0,\omega)$. Take the derivative

$$\frac{\partial}{\partial q} g(q\omega + v))|_{q=0} = \int_0^\infty \langle \omega, \nabla_v\rangle f(y+tv)t\mathrm{d}t.$$

Integrating against the element $\mathrm{d}\phi$ and setting $q=0$ we get

$$\begin{aligned}\int \frac{\partial}{\partial q} g(y, q\omega + rv)|_{q=0}\mathrm{d}\phi &= \int_0^{2\pi}\int_0^\infty \langle \omega, \nabla\rangle f(y+tv)t\mathrm{d}t\mathrm{d}\phi \\ &= \int_H \langle \omega, \nabla\rangle f\mathrm{d}S = \frac{\partial}{\partial p} Rf(H),\end{aligned}$$

since $\mathrm{d}S = t\mathrm{d}t\mathrm{d}\phi$ is the Euclidean surface density in H. ▶

Thus the quantity $\partial_p Rf(H)$ is reconstructed from data of ray integrals of f for the family of rays starting from a point $y \in H \cap \Gamma$ that is close to H. If the conditions of this theorem are satisfied for any hyperplane H in $\mathbf{E}$ that meets $\operatorname{supp} f$, we know the p-derivative of $Rf(H)$ for all H. The inversion formula for the Radon transform in the case $n=3$ depends only on the second p-derivative of Rf. Therefore the information we have is sufficient to apply this formula and recover the function f.

Corollary 4.7 *Let $x \in \mathbf{E}$ and $\Gamma \subset \mathbf{E}$ be a set such that each plane H through x meets Γ at a point $y(H)$. Then for an arbitrary $f \in C^2(\mathbf{E})$ with compact support disjoint with Γ, the reconstruction formula*

$$f(x) = -\frac{1}{8\pi^2}\int_{S^2} \mathrm{d}\omega \frac{\partial}{\partial p}\int_{\langle v,\omega\rangle=0} \frac{\partial}{\partial q} g(y(H), q\omega + rv(\phi))|_{q=0}\mathrm{d}\phi, \tag{4.10}$$

holds, where $L(u,y)$ and $\mathrm{d}\phi$ are as above.

◀ The integrand in the right side is equal to $\partial_p^2 Rf(H(\langle\omega, x\rangle, \omega))$. Then (4.10) follows from Radon's inversion formula (2.14). ▶

Generalization. Grangeat's method is generalized for arbitrary dimension n as follows:

Theorem 4.8 *Let H be a hyperplane in a Euclidean space $\mathbf{E}^n$, $y \in H$ and $f \in C^{n-2}(\mathbf{E})$ be a function such the set* $\operatorname{supp} f \cap H$ *is compact. Then we have*

$$\frac{\partial^{n-2}}{\partial p^{n-2}} Rf(H) = \sum_{k=0}^{[n/2]-1} c_{n,k}\int \frac{\partial^{n-2-2k}}{\partial q^{n-2-2k}} g(y, v + q\omega)|_{q=0}\mathrm{d}S,$$

where v runs over the unit sphere $\mathbf{S}^{n-2} \subset H - y$, $\mathrm{d}S$ is the area element on $\mathbf{S}^{n-2}$ and $c_{n,k} = (-1)^k((2k-1)!!)^2\binom{n-2}{2k}$.

◀ A proof can be done alogn the same lines as Theorem 4.6. ▶

4.4 Rays tangent to a surface

Theorem 4.9 *Let S be a smooth surface in* $\mathbf{E}$, *H be a plane in* $\mathbf{E}$ *transversal to S and K be a connected compact subset of $H\backslash S$. Then for an arbitrary $f \in C^2(\mathbf{E})$ with* $\operatorname{supp} f \cap H \subset K$, *we have*

$$\frac{\partial}{\partial p} Rf(H) = \frac{1}{z} \int_0^S \left[\frac{\kappa}{[x', \nu, \omega]} \frac{\partial}{\partial q} - \frac{\partial}{\partial s} \left(\frac{\langle \nu, \omega \rangle}{[x', \nu, \omega]} \right) \right] g(y, ry' + q\, y' \times \nu))|_{q=0} \mathrm{d}s, \tag{4.11}$$

where $y = y(s)$, $0 \le s \le S$ is the equation of the curve $\mathbf{C} \doteq S \cap H$ *such that $|y'| = 1$, $\kappa = [x', x'', \omega]$ is the curvature of* $\mathbf{C}$, *ω is a normal vector to H and $\nu = \nu(x(s))$ and $\nu(x)$ is a continuous unit normal field to S. We assume that the set $\xi^{-1}(K)$ is compact for the mapping*

$$\xi : \mathbf{C} \times \mathbb{R} \to H, \quad \xi(s,t) = y(s) + ty'(s)$$

and

$$z \doteq \sum_{x=\xi(s,t)} \operatorname{sgn} \kappa(s) \neq 0, \ x \in K.$$

Remark. The vectors $y'(s)$ and $y'(s) \times \nu$ are unit and belong to the pencil $\Sigma(S)$; the number z does not depend on $x \in K$.

◀ We have

$$\int\!\!\int \langle x', \nabla f \rangle \, t\mathrm{d}t = \int \frac{\partial}{\partial t} f(\xi)\mathrm{d}t = 0, \tag{4.12}$$

since the function $f|L(s,0)$ has compact support. Therefore

$$\begin{aligned} \frac{\partial}{\partial s} g(y(s), y'(s)) &= \int \frac{\partial}{\partial s} f(\xi(s,t))\mathrm{d}t = \int \langle \nabla f(\xi), y'(s) + ty''(s) \rangle \, \mathrm{d}t \\ &= \int \langle \nabla f(\xi), y''(s) \rangle \, t\mathrm{d}t = \kappa \int [\nabla f(\xi), \omega, y'(s)] \, t\mathrm{d}t \end{aligned} \tag{4.13}$$

because of $y'' = \kappa\omega \times y'$. Further

$$\frac{\partial}{\partial q} g(y(s), ry'(s) + qy'(s) \times \nu)|_{q=0} = \int [\nabla f(\xi), y'(s), \nu] \, t\mathrm{d}t.$$

The vectors ω, $y' \times \omega$, $y' \times \nu$ are orthogonal to y' hence $y' \times \nu = [y', \nu, \omega]\omega + \langle \nu, \omega \rangle \, y' \times \omega$. This yields

$$\frac{\kappa}{[x', \nu, \omega]} \frac{\partial}{\partial q} g(y, ry' + qy' \times \nu)|_{q=0} = \kappa \int (f'(\xi), \omega) \, t\mathrm{d}t + \frac{\langle \nu, \omega \rangle}{[y', \nu, \omega]} \frac{\partial}{\partial s} g(y, y').$$

Integrate both sides over $H \cap S$ against the density $\mathrm{d}s$:

$$\begin{aligned} &\int \frac{\kappa}{[x', \nu, \omega]} \frac{\partial}{\partial q} g(y, y')\,\mathrm{d}s + \int \frac{\langle \nu, \omega \rangle}{[x', \nu, \omega]} \frac{\partial}{\partial s} g(y, y')\,\mathrm{d}s \\ &\quad = \int \kappa \int \langle \nabla f(\xi), \omega \rangle \, t\mathrm{d}t\mathrm{d}s = \frac{\partial}{\partial \omega} \int \kappa \int f(\xi) t\mathrm{d}t\mathrm{d}s. \end{aligned} \tag{4.14}$$

Consider the system of coordinates $s, t, p = \langle \omega, x \rangle$ in a neighborhood of $\mathbf{C}$. We have $\partial x/\partial(t, s, p) = \det(y', y' + ty'', \omega) = \kappa t$. The Euclidean volume density in H is equal to $\mathrm{d}V/\mathrm{d}p$, hence $\partial h/\partial(s, t) = \kappa t$. Therefore

$$\int \kappa \int f(\xi) t \mathrm{d}t \mathrm{d}s = z \int_H f \mathrm{d}S = z \, Rf(H)$$

hence, the right side of (4.14) is equal to $z \, \partial Rf(H)/\partial p$ and

$$z \frac{\partial}{\partial p} Rf(H) = \int \frac{\kappa}{[x', \nu, \omega]} \frac{\partial}{\partial q} g(y, y') \, \mathrm{d}s + \int \frac{\langle \nu, \omega \rangle}{[x', \nu, \omega]} \frac{\partial}{\partial s} g(y, y') \, \mathrm{d}s.$$

Integrating by parts in the second term, we get (4.11). ▶

Remark 1. Theorem (4.6) is a limiting case of the above result. Indeed, take an arbitrary compact smooth curve $\Gamma \subset \mathbf{V}$ and consider its ε-neighborhood Γ_ε for some $\varepsilon > 0$. If the number ε is sufficiently small, the boundary S_ε of Γ_ε is a smooth surface. Take the pencil $\mathbf{\Sigma}(S_\varepsilon)$ of tangent lines and apply formula (4.11). It is easy to see that this pencil tends to the pencil $\mathbf{\Sigma}(\Gamma)$ as $\varepsilon \to 0$ and formula (4.11) tends to (4.9).

Remark 2. The case $n > 3$ is more complicated since the affine transform M_k fulfils a complicated system of differential equations. Anyway we obtain in Chapter 5 a reconstruction method for the case $k = n - 2$ by means of duality arguments.

4.5 Sources on a proper curve

The formula (4.10) contains a three-fold integral. The following result gives a reconstruction in Euclidean space $\mathbf{E}$ of arbitrary dimension n by means of a two-fold integration.

Theorem 4.10 *Let Γ be an oriented C^1-curve in $\mathbf{E}$ joining some points a and b. Then for an arbitrary function $f \in C^2(\mathbf{E})$ such that* $\operatorname{supp} f \Subset \mathbf{E} \backslash \mathbf{\Gamma}$ *and any point $x \in]a, b[$ the equation*

$$2\pi^2 f(x) = -\int_{y \in \mathbf{\Gamma}} \frac{\mathrm{d}\gamma}{|y - x|} \int \frac{\partial}{\partial \phi} g(y, v(\phi)) \frac{\mathrm{d}\phi}{\sin \phi} \tag{4.15}$$

$$- \int_{y \in \mathbf{\Gamma}} \mathrm{d} \frac{1}{|y - x|} \int g(y, v(\phi)) \frac{\mathrm{d}\phi}{\sin \phi} + \frac{1}{|y - x|} \int g(y, v(\phi)) \frac{\mathrm{d}\phi}{\sin \phi} \Big\lceil_{y=a}^{y=b},$$

holds, where $\{y = y(s), 0 \le s \le 1\}$ is a parametrization of Γ, $\mathrm{d}\gamma$ is the angle a piece of $\mathbf{\Gamma}$ is seen from x, $v = v(\phi)$ is a unit vector in the plane spanned by $y - x$ and the tangent vector y' to Γ at y with the polar angle ϕ which is specified by the conditions $\phi(y - x) = 0$, $\sin \phi(y') \ge 0$.

The geometry is shown in Fig. 4.1. The integral against the density $\mathrm{d}\phi/\sin\phi$ is taken over the interval $[0, 2\pi]$ as principle value. The set of points s where the vectors $y - x$ and y' are collinear, is omitted in the exterior (it has zero measure). The inner integral in the first term does not depend on the orientation of $\mathbf{\Gamma}$ and so does the density $\mathrm{d}\gamma$. The same is true for the second and the third terms, since changing the orientation of $\mathbf{\Gamma}$ multiplies by -1 for the inner integral.

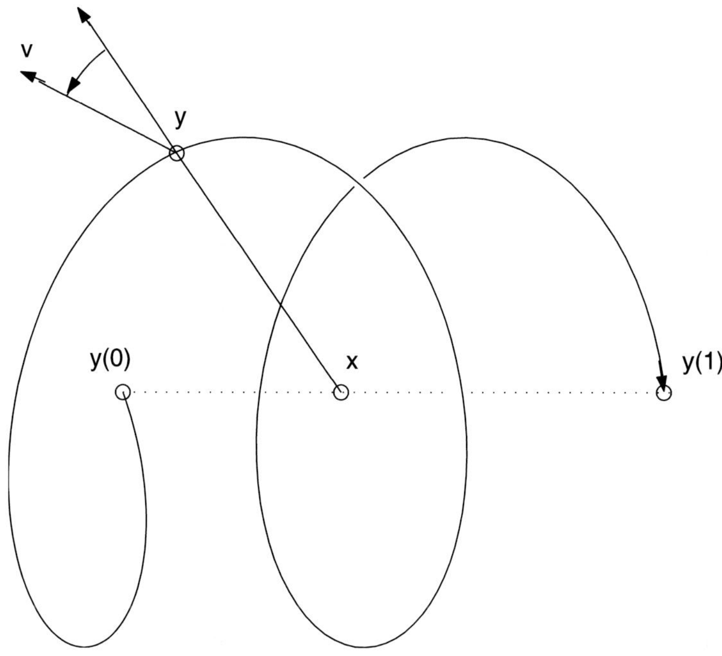

Figure 4.1: Rays with vertices on the curve

◀ For an arbitrary point $y \neq x$ in $\mathbf{E}$, an oriented plane T that contains x and y and a tangent vector θ in T we consider the integral

$$I(y) \doteq \int_0^{2\pi} g(y, v(\phi)) \frac{\mathrm{d}\phi}{\sin\phi},$$

where $v(\phi)$ is the unit tangent vector in T whose angle with θ is equal to ϕ. Take the unit tangent vector η in T that is orthogonal to θ such that the pair (θ, η) defines the orientation of T. Note that $G(y, \theta) \doteq g(y, v(0)) + g(y, v(\pi))$ is the line integral of f.

Lemma 4.11 *We have*

$$I(y) = \int G(y + q\eta, \theta) \frac{\mathrm{d}q}{q}.$$

◀ Substituting (4.8) and denoting $q = r \sin\phi$, gives

$$I(y) \doteq \int_0^{2\pi} \int_{L(y,v(\phi))} f \, \mathrm{d}r \frac{\mathrm{d}\phi}{\sin\phi} = \int_T f \, \mathrm{d}r \frac{\mathrm{d}\phi}{\sin\phi} = \int_{\mathbb{R}} \int_{\mathbb{R}} f\,(y + p\theta + q\eta) \, \mathrm{d}p \frac{\mathrm{d}q}{q}$$
$$= \int G(y + p\eta, \theta) \frac{\mathrm{d}p}{p}.$$ ▶

By the lemma, $\langle y, \nabla_y \rangle I(y) = 0$ and

$$\langle \eta, \nabla_y \rangle I(y) = \int \frac{\partial}{\partial q} G(y + q\eta, \theta(y)) \frac{\mathrm{d}q}{q}.$$

Choose a natural parameter s in $\mathbf{\Gamma}$ and integrate both sides along $\mathbf{\Gamma}$:

$$\int_{\mathbf{\Gamma}} \langle \eta, \nabla_y \rangle I(y) \, \mathrm{d}\gamma = \int_{\mathbf{\Gamma}} \mathrm{d}\gamma \int \frac{\partial}{\partial q} G(y + q\eta, \theta(y)) \frac{\mathrm{d}q}{q}. \tag{4.16}$$

Take the projection $\mathbf{\Gamma}(\infty)$ of $\mathbf{\Gamma}$ to the unit sphere $\mathbf{S}$ centred at x. The projections $a(\infty)$ and $b(\infty)$ of the endpoints are opposite in $\mathbf{S}$. The function $G'_q(y + q\eta, \theta(y))$ does not change, if we move the point $y \in \mathbf{\Gamma}$ parallel to the vector $\theta(y)$. Therefore we can replace the curve $\mathbf{\Gamma}$ by $\mathbf{\Gamma}(\infty)$ in this integral. Now we can apply Theorem 4.3. By (4.3) the right-hand side is equal to $-2\pi^2 f(x)$. Take a parametrization $y = y(s)$ of the curve $\mathbf{\Gamma}$ and set $\theta = |y - x|^{-1}(y - x)$. If the vector $y'(s)$ is not collinear to θ the pair (θ, y') defines an orientation of T. Choose a unit vector η in T orthogonal to θ that is consistent with this orientation. By the lemma we have

$$\langle \eta, \nabla_y \rangle I(y) \, \mathrm{d}\gamma = \frac{\mathrm{d}s}{|y - x|} \langle y'(s), \nabla_y \rangle I(y)$$
$$= \frac{\mathrm{d}s}{|y - x|} \int_0^{2\pi} \langle y'(s), \nabla_y \rangle g(y(s), v(s, \phi)) \frac{\mathrm{d}\phi}{\sin\phi},$$

since $\eta \mathrm{d}\gamma$ equals the projection of the covector $|y - x|^{-1} y'(s) \, \mathrm{d}s$ to $\mathbf{\Gamma}(\infty)$. Further

$$\langle y'(s), \nabla \rangle g(y(s), v(s, \phi)) = \frac{\partial}{\partial s} g(y(s), v(s, \phi)) - \frac{\mathrm{d}\gamma}{\mathrm{d}s} \frac{\partial}{\partial \phi} g(y(s), v(s, \phi)),$$

which yields

$$\int_{\mathbf{\Gamma}} \langle \eta, \nabla_y \rangle I(y) \, \mathrm{d}\gamma = \int_{\mathbf{\Gamma}} \frac{\mathrm{d}s}{|y - x|} \langle y', \nabla_y \rangle I(y)$$
$$= \int_{\mathbf{\Gamma}} \frac{\mathrm{d}s}{|y - x|} \frac{\partial}{\partial s} \int_0^{2\pi} g(y(s), v(s, \phi)) \frac{\mathrm{d}\phi}{\sin\phi}$$
$$- \int_{\mathbf{\Gamma}} \frac{\mathrm{d}\gamma}{|y - x|} \int_0^{2\pi} \frac{\partial}{\partial \phi} g(y(s), v(s, \phi)) \frac{\mathrm{d}\phi}{\sin\phi}.$$

Integrating by parts in the first term, we get the right side of (4.15). Comparing with (4.16) completes the proof of Theorem 4.10. ▶

4.6 Reconstruction from plane integrals of sources

Let $\varphi = f\mathrm{d}x$ be a density of sources of radiation in the Euclidean space $\mathbf{E}^3$ with compact support. For a plane $P \subset \mathbf{E}$ the plane density of sources in P is equal to $\varphi/\mathrm{d}l$ where l is a linear function in $\mathbf{E}$ of the norm 1 that vanishes on P. We can write $\varphi/\mathrm{d}l = f\,\mathrm{d}S$ where $\mathrm{d}S$ is the area element in P. For a point $a \in P$ the total of the radiation from sources in P, measured by a detector in position a, is given by the integral

$$I(a, P; \varphi) \doteq \int_P \frac{f(x)\mathrm{d}S}{|x-a|}.$$

The problem is to reconstruct f from knowledge of integrals $I(a, P)$ for a set of sources a and planes P. Suppose that these integrals are known for all sources a on a C^1-curve Γ in $\mathbf{E}$ and all planes P through each source a. Denote by K the convex hull of $\operatorname{supp}\varphi$.

Theorem 4.12 *Let Γ be a smooth curve in $\mathbf{E}\backslash K$ that fulfils the completeness condition: any plane P that meets* $\operatorname{supp}\varphi$ *contains a point $a \in \Gamma$. The density φ can be reconstructed from the knowledge of $I(a, P; \varphi)$ for all $a \in \Gamma$ and $P \ni a$.*

◀ Take a point $a \in \Gamma$ and choose a Euclidean coordinate system y_1, y_2, y_3 centred at a such that $y_1 > 0$ in $\operatorname{supp}\varphi$. Take a plane P through a that touches $\operatorname{supp}\varphi$ and rotate the coordinates (y_2, y_3) in such a way that $P = \{y_3 = wy_1\}$. We have $\mathrm{d}S = \sqrt{1+w^2}\mathrm{d}y_1\mathrm{d}y_2$ and

$$\frac{1}{\sqrt{1+w^2}} I(a, P(w); \varphi) = \int_P \frac{f\,\mathrm{d}y_1\mathrm{d}y_2}{|x-a|}.$$

The projective transformation

$$u = \frac{1}{y_1},\ v = \frac{y_2}{y_1},\ w = \frac{y_3}{y_1},\ \mathrm{d}y_1\mathrm{d}y_2 = u^{-3}\mathrm{d}u\mathrm{d}v$$

yields the equations $|x-a| = u^{-1}\sqrt{1+v^2+w^2}$,

$$\frac{\mathrm{d}y_1\mathrm{d}y_2}{|x-a|} = \frac{\mathrm{d}u\mathrm{d}v}{u^2\sqrt{1+v^2+w^2}}$$

and

$$\begin{aligned}\int_P \frac{f\,\mathrm{d}y_1\mathrm{d}y_2}{|x-a|} &= \int_0^\infty \frac{f(y_1, vy_1, wy_1)\mathrm{d}u}{u^2} \frac{\mathrm{d}v}{\sqrt{1+v^2+w^2}} \qquad (4.17)\\ &= \int_{-\infty}^{\infty}\int_{r(v,w)} f\mathrm{d}y_1 \frac{\mathrm{d}v}{\sqrt{1+v^2+w^2}},\end{aligned}$$

where $r(v, w) \doteq \{y_2 = vy_1, y_3 = wy_1,\ y_1 \geq 0\}$. Denote

$$Lf(v, w) \doteq \frac{1}{\sqrt{1+v^2+w^2}} \int_{r(v,w)} f\,\mathrm{d}y_1 = \frac{1}{1+v^2+w^2} \int_{r(v,w)} f\mathrm{d}s, \qquad (4.18)$$

where $\mathrm{d}s$ is the line density in the ray $r(v,w)$. Consider (4.18) as a function in (v,w)-plane. The equation (4.17) yields

$$\int Lf(v,w)\mathrm{d}v = \frac{I(a,P;\varphi)}{\sqrt{1+w^2}},$$

which means that we know the line integral of this function over any line $w = \text{const}$. Any rotation in the (v,w)-plane is generated by a rotation of the plane P around the y_1-axis. Therefore we know all line integrals of $L(v,w)$ and recover this function by means of any inversion method for the plane Radon transform. From (4.18), we know the ray transform $g(r)$ of the density function f for any ray $r = r(v,w)$ with the source a. When the point a runs over Γ, we get data of all integrals over rays with sources in Γ and the function f can be recovered by means of Theorem 4.6 or of Theorem 4.10. ▶

4.7 Line integrals of differential forms

Let now α be a differential form of degree 1 in a vector space $\mathbf{V}^n$. Choose a coordinate system in $\mathbf{V}$ and write $\alpha = \sum a_j(x)\,\mathrm{d}x_j$. Suppose that the coefficients $a_1,\ldots,a_n$ are integrable in each affine (straight) line $L \subset \mathbf{V}$. Choose an orientation of a line L and consider the integral

$$\int_L \alpha = g(x,\theta) \doteq \sum_{j=1}^{n} \int_{-\infty}^{\infty} a_j(x+t\theta)\,\theta_j \mathrm{d}t$$

where $y = x + t\theta$ is a parameterization of L where $x,\theta \in \mathbf{V}$. We can replace x by $x' = x + s\theta$ for any $s \in \mathbb{R}$ without changing the integral; (therefore the parameter x can be restricted to an $n-1$-subspace $\theta^{\perp}$ that does not contain θ). We call the function $I\alpha(L) = \int_L \alpha$ defined on the variety $\mathsf{A}_1^*(\mathbf{V})$ of all oriented affine lines, a *line transform* of the form α. We, of course, have $I\alpha(L') = -I\alpha(L)$ for the opposite orientation L' of the line L.

We cannot expect that the form α can be reconstructed from $I\alpha$, since all the line integrals vanish for any form $\alpha = \mathrm{d}\phi$, where ϕ is a C^1-function vanishing at infinity. It is easy to check the inverse statement:

Proposition 4.13 *If $I\alpha(L) = 0$ for all lines $L \subset \mathbf{V}$ and a form α as above, then $\alpha = \mathrm{d}\phi$ for a function ϕ that vanishes at infinity.*

◀ Apply induction in $n = \dim \mathbf{V}$. In the case $n = 1$ we take the primitive

$$\phi(x) = \int_{-\infty}^{x} \alpha$$

which vanishes at $-\infty$ and also at ∞ due to the condition $I\alpha(\mathbf{V}) = 0$. In the case $n > 1$ we apply the inductive assumption to the form $\alpha' = \sum_1^{n-1} a_j \mathrm{d}x_j$ and to each affine subspace $\mathbf{V}(s) = \{x_n = s\}$ and get a function ψ vanishing at infinity such that $\partial\psi/\partial x_j = a_j$, $j = 1, \dots, n-1$. The function ψ is uniquely defined. Consider the form $\beta = \alpha - \mathrm{d}\psi$; we have $\beta = b\mathrm{d}x_n$, where the function b is integrable in each affine line. We again have $I\beta(L) = 0$, which immediately implies $b = 0$. ▶

On the other hand, the equation $\mathrm{d}\alpha = 0$ is also equivalent to the representation $\alpha = \mathrm{d}\phi$ which means that we cannot reconstruct more than the differential $\mathrm{d}\alpha$ from knowledge of $I\alpha$. Therefore the reconstruction problem is stated as follows: how many line integrals do we need to reconstruct the form $\mathrm{d}\alpha$? The manifold $\mathsf{A}_1^*(\mathbf{V})$ has dimension $2n - 2$, hence the data of all line integrals is redundant. We will choose an n-dimensional submanifold Σ (pencil). Following the approach of Section 4.2 we take a closed homotopicaly non-trivial curve $\mathbf{C}$ in $\mathbb{P}(\mathbf{V})$ and consider the pencil $\Sigma = \Sigma(\mathbf{C})$ of affine lines in $\mathbf{V}$ that meets at infinity the curve $\mathbf{C}$. In the case $n = 2$ this pencil coincides with the variety of all affine lines.

Proposition 4.14 *Let $n = 2$ and the form $\mathrm{d}\alpha$ be integrable in $\mathbf{V}$. This form can be explicitly reconstructed from the line transform of α.*

◀ Take an oriented line L and the half-plane U such that $\partial U = L$. By the Green formula $I\alpha(L) = \int_U \mathrm{d}\alpha$. Introduce the Euclidean structure by means of coordinates x_1, x_2 in the plane $\mathbf{V}$ and write $\mathrm{d}\alpha = a\,\mathrm{d}x_1 \wedge \mathrm{d}x_2$ for a function a. We know the integrals of the function a over any half-plane. Taking derivatives in the normal direction yields line integrals of the function a. We can reconstruct this function by applying the Radon inversion formula. ▶

If $n > 2$, it is not possible to reconstruct $\mathrm{d}\alpha$ from the line transform of α known on the pencil $\Sigma(\mathbf{C})$ even for a homotopically non-trivial curve $\mathbf{C}$ (unlike the situation of Section 4.2).

Example 1. Take for $\mathbf{C}$ any projective line, for instance, the projective line $\mathbf{L}$ at the plane coordinate (x_1, x_2). The pencil $\Sigma(\mathbf{L})$ consists of all lines parallel to this plane. The form $\alpha = \sum_3^n a_j \mathrm{d}x_j$ with arbitrary coefficients $a_3, \dots, a_n$ has zero integrals over all such lines.

To have more information, we take $n-1$ closed curves $\mathbf{C}_1, \dots, \mathbf{C}_{n-1}$ in $\mathbb{P}(\mathbf{V})$ and set $\mathbf{C} = \cup\mathbf{C}_j$. It appears that the pencil $\Sigma(\mathbf{C})$ is big enough for reconstruction of $\mathrm{d}\alpha$ provided the curves $\mathbf{C}_j$ are in general position and each of them is non-trivial, i.e., is not homotopic to a point.

The case $n = 3$. Choose a coordinate system x_1, x_2, x_3 in a three dimensional space $\mathbf{V}$ and define the orientation and the Euclidean structure by means of these coordinates. Denote for convenience $g(x, \theta) = I\alpha(L(x, \theta))$ the integral of form α over the line $L(x, \theta)$ through the point x parallel to the unit vector θ. Take some smooth closed non-trivial curves $\mathbf{C}_1, \mathbf{C}_2$ in $\mathbb{P}(\mathbf{V})$. We choose smooth parameterizations $\theta_j = \theta_j(s_j)$, $0 \le s_j \le S_j$ of $\mathbf{C}_j$ such that $|\theta_j| = |\theta_j'| = 1$, $\theta_j(S_j) = -\theta_j(0)$, $j = 1, 2$.

Theorem 4.15 *Let $\mathbf{C}_1, \mathbf{C}_2 \subset \mathbb{P}(\mathbf{V}^3)$ be C^1-smooth closed non trivial C^1-curves such that the set $\mathbf{C}_1 \cap \mathbf{C}_2$ is finite. For any 1-form α of class C^3 with compact support in $\mathbf{V}$ the equation*

$$4\pi^2 \mathrm{d}\alpha(x) = \int_0^{S_1} \mathrm{d}s_1 \int_0^{S_2} \mathrm{d}s_2 \left\langle \frac{\theta_1 \times \theta_2}{|\theta_1 \times \theta_2|}, \nabla_x \right\rangle \langle \theta_2', \nabla_x \rangle \langle \theta_1', \nabla_x \rangle G(x; s_1, s_2) \vdash \mathrm{d}x, \tag{4.19}$$

$$G(x; s_1, s_2) \doteq \left(\int_{-\infty}^{\infty} g(x + t\theta_1, \theta_2) \,\mathrm{d}t\ \theta_1 - \int_{-\infty}^{\infty} g(x + t\theta_2, \theta_1) \,\mathrm{d}t\ \theta_2 \right)$$

gives a reconstruction of the form $\mathrm{d}\alpha$ from integrals of α over lines $L \in \Sigma(\mathbf{C}_1 \cup \mathbf{C}_2)$.

Remark 1. The coefficient $|\theta_1 \times \theta_2| = 0$ vanishes at any point (s_1, s_2), where $\theta_1(s_1) \neq \pm\theta_2(s_2)$, at any common point of $\mathbf{C}_1$ and $\mathbf{C}_2$. There is no singularity at this point, since the vector $\omega = |\theta_1 \times \theta_2|^{-1} \theta_1 \times \theta_2$ stays bounded. The set of common points of the curves is finite by assumption. On the other hand, any two non-trivial closed curves in the projective plane have, at least, one common point.

Remark 2. The form $G(x; s_1, s_2)$ depends only on $p \doteq \langle \omega, x \rangle$ and the derivative $\langle \theta_i', \nabla_x G \rangle$ can be written as $\langle \theta_i', \omega \rangle \langle \omega, \nabla_x \rangle G$, $i = 1, 2$. The alternative form of (4.19) looks as follows

$$4\pi^2 \mathrm{d}\alpha(x) = \int_0^{S_1} \langle \mathrm{d}\theta_1, \omega \rangle \int_0^{S_2} \langle \mathrm{d}\theta_2, \omega \rangle \langle \omega, \nabla_x \rangle^3 G(x; s_1, s_2) \vdash \mathrm{d}x.$$

◄ Proof of Theorem 4.15. Let $\beta = b^{ij} \mathrm{d}x_i \wedge \mathrm{d}x_j$ be a two-form in $\mathbf{V}$ and A an operator defined on functions in $\mathbf{V}$; we shall use the notation $A[\beta] = \sum A[b^{ij}] \mathrm{d}x_i \wedge \mathrm{d}x_j$. In particular, the notation $\widehat{[\mathrm{d}\alpha]}$ means the Fourier transform applied to each coefficient of $\mathrm{d}\alpha$. By the "slice" Theorem

$$\widehat{[\mathrm{d}\alpha]}(\xi) = \int_{\theta^\perp} \exp(-\mathrm{j}p|\xi|) R[\mathrm{d}\alpha] \left(p, |\xi|^{-1} \xi\right) \mathrm{d}p. \tag{4.20}$$

It is easy to evaluate the Radon transform of the form from λ. Take some unit vectors $\theta_j = \theta_j(s_j) \in \mathbf{C}_j, j = 1, 2$ that are orthogonal to ξ and choose the functionals $y_1, y_2 \in \mathbf{V}^*$ such that $y_i(\theta_j) = \delta_{ij}$, $y_j(\omega) = 0$. The functions $y_1 = y_1(x), y_2 = y_2(x), y_3 = \omega(x)$ form a coordinate system in $\mathbf{V}$.

Lemma 4.16 *We have for any unit vector ω and point x,*

$$R[\mathrm{d}\alpha](\langle \omega, x \rangle, \omega) = |\sin\phi| \langle \omega, \nabla_x \rangle G(x, s_1, s_2), \tag{4.21}$$

where ϕ is the angle between θ_1 and θ_2.

◄ Express the coordinates y through x :

$$y_3 = \langle \omega, x \rangle, \quad y_1 = \sin^{-2}\phi \left(\langle \theta_1, x \rangle - \cos\phi \langle \theta_2, x \rangle \right),$$
$$y_2 = \sin^{-2}\phi \left(\langle \theta_2, x \rangle - \cos\phi \langle \theta_1, x \rangle \right)$$

We can write the unknown form in the coordinates $y : \alpha = a\mathrm{d}y_1 + b\mathrm{d}y_2 + c\mathrm{d}y_3$ and

$$\mathrm{d}\alpha = (b_1' - a_2')\,\mathrm{d}y_1 \wedge \mathrm{d}y_2 + (c_2' - b_3')\,\mathrm{d}y_2 \wedge \mathrm{d}y_3 + (a_3' - c_1')\,\mathrm{d}y_3 \wedge \mathrm{d}y_1$$

where $a_j' = \partial a/\partial y_j$, $j = 1, 2, 3$.

Integrate by means of the equation $\mathrm{d}S = |\sin\phi|\,\mathrm{d}y_1\mathrm{d}y_2$:

$$\int_{H(p,\omega)} (a_3' - c_1')\,\mathrm{d}S = |\sin\phi| \int_H a_3'\mathrm{d}y_1\mathrm{d}y_2 = |\sin\phi| \int \mathrm{d}y_2 \int a_3'\mathrm{d}y_1.$$

The inner integral is taken over a line L parallel to the vector θ_1, hence $L \in \Sigma(\mathbf{C})$ and the integral equals $g_3'(y_2, p, \theta_1)$. Similarly, we have

$$\int_{H(p,\omega)} (c_2' - b_3')\,\mathrm{d}S = |\sin\phi| \int \mathrm{d}y_1 g_3'(y_1, p, \theta_2),$$

and obviously $\int_H (b_1' - a_2')\,\mathrm{d}S = 0$. Returning to the coordinates x_j yields (4.21). ▶

Further, we follow the arguments of Section 4.2 with few modifications. Let $\mathbf{S}$ be the unit sphere in $\mathbf{V}$; we have the natural surjection $\mathbf{p} : \mathbf{S} \to \mathbb{P}(\mathbf{V})$, which glues together each pair of opposite points of the sphere. Take the closed curve $\mathbf{D}_1 \subset \mathbf{S}$ that covers $\mathbf{C}_1$ twice with the parameterization $\theta = \theta_1(s_1)$ $0 \le s_1 \le 2S_1$ such that $\theta_1(s + S_1) = -\theta_1(s), \theta_2(2S_2) = \theta_2(0)$. Let $\mathbf{D}_2, \theta_2, S_2$ have a similar meaning for the curve $\mathbf{C}_2$ and $\mathbf{D} = \mathbf{D}_1 \times \mathbf{D}_2$.

Let $\xi = (\xi_1, \xi_2, \xi_3)$ be the dual coordinate system in the dual space $\mathbf{V}^*$; consider the volume form $\upsilon \doteq \mathrm{d}\xi_1 \wedge \mathrm{d}\xi_2 \wedge \mathrm{d}\xi_3$. Consider the 5-manifold $\mathbf{D} \times \mathbf{V}^*$ oriented by the form $\upsilon \wedge \mathrm{d}s$, where $\mathrm{d}s = \mathrm{d}s_1 \wedge \mathrm{d}s_2$. The restriction of the form $\mathrm{d}\langle\theta_j(s_j), \xi\rangle$ to $\mathbf{D} \times \mathbf{V}^*$ is equal to $\langle\theta_j(s_j), \mathrm{d}\xi\rangle + \langle\theta_j'(s_j), \xi\rangle\,\mathrm{d}s_j, j = 1, 2$ and does not vanish. Denote $\Theta = \Theta(s, \xi) \doteq \langle\theta_1'(s_1), \xi\rangle\langle\theta_2'(s_2), \xi\rangle$. The differential 3-form $\eta = \Theta^{-1}\upsilon$ on $\mathbf{D} \times \mathbf{V}^*$ fulfils the equation $\eta \wedge \mathrm{d}\langle\theta_1, \xi\rangle \wedge \mathrm{d}\langle\theta_2, \xi\rangle = \upsilon \wedge \mathrm{d}s$. Introduce the submanifold

$$\Gamma \doteq \{(s, \xi) : \langle\theta_1(s_1), \xi\rangle = \langle\theta_2(s_2), \xi\rangle = 0,\ \Theta(s, \xi) \ge 0\}$$

and define the orientation on Γ by means of the form η. Consider the mapping $\gamma_{\mathbf{C}} : \Gamma \to \mathbf{V}^*$, $\gamma_{\mathbf{C}}(s, \xi) = \xi$. The pair $\Pi = (\Gamma, \gamma_{\mathbf{C}})$ is an odd n-chain in $\mathbf{V}^*$.

Lemma 4.17 *The chain Π is closed.*

◀ For any $s = (s_1, s_2)$ the half-line $\Gamma(s) \doteq \{\xi : (s, \xi) \in \Gamma\}$ is complementary to $\Gamma(s + S)$ and their orientations are given by the same form η. This implies that the boundaries of these half-lines have opposite orientation and the lemma follows. ▶

The mapping $\gamma_{\mathbf{C}}$ is proper and its topological degree $\deg\gamma_{\mathbf{C}}$ is well defined.

Lemma 4.18 $\deg\gamma_{\mathbf{C}} = 1$.

◀ Denote this mapping by $\gamma_{\mathbf{C}}$. The degree of $\gamma_{\mathbf{C}}$ is invariant under any deformation of the curves $\mathbf{C}_j, j = 1, 2$. We have a homotopy $\mathbf{C}_1 \sim \mathbf{L}_1, \mathbf{C}_2 \sim \mathbf{L}_2$ in this class where $\mathbf{L}_1, \mathbf{L}_2$ are different projective lines in $\mathbb{P}(\mathbf{V})$. Therefore we have $\deg \gamma_{\mathbf{C}} = \deg \gamma_{\mathbf{L}}$. The mapping $\gamma_{\mathbf{L}}$ restricted to the interior of Γ is a bijection onto $\mathbf{V}^* \backslash \mathbf{L}^*$ where $\mathbf{L}^*$ is the orthogonal complement to the line $\mathbf{L}_1 \cap \mathbf{L}_2$. The orientation of Γ defined by η coincides with the orientation of $\mathbf{V}$ since $\Theta > 0$ in Γ. This yields the equation $\deg \gamma_{\mathbf{L}} = 1$. ▶

Lemma 4.18 implies that

$$\mathrm{d}\alpha(x) = \int_{\mathbf{E}} \exp(\mathrm{j} \langle \xi, x \rangle) \widehat{\lceil \mathrm{d}\alpha \rceil}(\xi)\, v = \int_{\Gamma} \exp(\mathrm{j} \langle \xi, x \rangle) \widehat{\lceil \mathrm{d}\alpha \rceil}(\xi) \gamma^* (v)\,, \tag{4.22}$$

where $\gamma^* (v)$ is the pull back of v under the mapping $\gamma = \gamma_{\mathbf{C}}$. Take a 2-form ψ in $\mathbf{D} \times \mathbf{V}^*$ that satisfies $\psi \wedge \langle \theta_1, \mathrm{d}\xi \rangle \wedge \langle \theta_2, \mathrm{d}\xi \rangle = v$ and set $\chi = \Theta\, \psi \wedge \mathrm{d}s$. We claim that $\gamma^* (v) = \chi$ in Γ. Indeed, it is easy to check thet the form $\rho \doteq v - \chi$ defined in $\mathbf{D} \times \mathbf{E}^*$ fulfils the equation $\rho \wedge \mathrm{d} \langle \theta_1, \xi \rangle \wedge \mathrm{d} \langle \theta_2, \xi \rangle = 0$. This implies that $\rho = \sigma_1 \wedge \mathrm{d} \langle \theta_1, \xi \rangle + \sigma_2 \wedge \mathrm{d} \langle \theta_2, \xi \rangle$, where both terms vanish in Γ. Therefore we can replace $\gamma^* (v)$ by χ in (4.22); substituting (4.20) gives the integral

$$\begin{aligned} &\int_{\Gamma} \exp(\mathrm{j} \langle \xi, x \rangle) \widehat{\lceil \mathrm{d}\alpha \rceil}(\xi)\, \gamma^* (v) = \int_{\Gamma} \exp(\mathrm{j} \langle \xi, x \rangle) \widehat{\lceil \mathrm{d}\alpha \rceil}(\xi)\, \chi \\ &= \int_0^{2S_1} \mathrm{d}s_1 \int_0^{2S_2} \mathrm{d}s_2 \int_{\Gamma(s)} \Theta(s, \xi) \exp(\mathrm{j} \langle \xi, x - y \rangle)\, \psi\, R \lceil \mathrm{d}\alpha \rceil (\langle \omega, y \rangle, \omega)\,, \end{aligned} \tag{4.23}$$

where $\omega = \xi / |\xi|$. For any fixed $s_j \in [0, 2S_j]$ the form ψ defines the orientation of the ray $\Gamma(s)$ and $\psi = \mathrm{d}p / |\sin \phi|$. We have $y - x = p\omega + z$, where $p = \langle \omega, y - x \rangle$, $\langle \omega, z \rangle = 0$, which yields $\langle \xi, x - y \rangle = -p |\xi|$. Substituting in the inner integral yields

$$\begin{aligned} \int_{\Gamma(s)} \Theta \exp(\mathrm{j} \langle \xi, x - y \rangle) \psi &= \frac{\langle \theta_1', \omega \rangle \langle \theta_2', \omega \rangle}{|\sin \phi|} \int_0^{\infty} |\xi|^2 \exp(-\mathrm{j} p |\xi|)\, \mathrm{d}|\xi| \\ &= \frac{2 \langle \theta_1', \omega \rangle \langle \theta_2', \omega \rangle}{|\sin \phi|} (\mathrm{j}(p - 0\imath))^{-3}. \end{aligned}$$

Substituting in (4.23) gives the quantity

$$\int_0^{2S_1} \langle \mathrm{d}\theta_1, \omega \rangle \int_0^{2S_2} \langle \mathrm{d}\theta_2, \omega \rangle \frac{1}{|\sin \phi|} \int_{\mathbb{R}} \frac{R \lceil \mathrm{d}\alpha \rceil (\langle \omega, y \rangle, \omega)}{(\mathrm{j}(p - 0\imath))^3}\, \mathrm{d}p.$$

Sum up the contributions of points s_1 and $s_1 + S_1$ and of points s_2 and of $s_2 + S_2$ for $0 \le s_j \le S_j, j = 1, 2$ using the relations

$$\begin{gathered} \omega(s_1 + S_1, s_2) = \omega(s_1, s_2 + S_2) = -\omega(s_1, s_2) = -\omega(s_1 + S_1, s_2 + S_2)\,, \\ (\mathrm{j}(p - 0\imath))^{-3} + (\mathrm{j}(-p - 0\imath))^{-3} = -\frac{1}{8\pi^2} \delta^{(2)}(p)\,. \end{gathered}$$

This results in the integral

$$-\frac{1}{4\pi^2}\int_0^{S_1}\langle \mathrm{d}\theta_1,\omega\rangle\int_0^{S_2}\langle \mathrm{d}\theta_2,\omega\rangle\frac{1}{|\sin\phi|}\langle\omega,\nabla\rangle^2 R\lceil \mathrm{d}\alpha\rceil(\langle\omega,x\rangle,\omega).$$

Applying Lemma 4.16 yields (4.19). ▶

Remark 3. An arbitrary 1-form α in a Euclidean space $\mathbf{E}^n$ can be represented as the sum $\alpha=\beta+\mathrm{d}\phi$, where ϕ is a function and the form β fulfils the equation $\mathrm{d}^*\beta=0$ where d^* is the adjoint operator to d. It is called the Helmholtz decomposition and a form β satisfying this equation is called a solenoidal form. If the form $\gamma=\mathrm{d}\alpha$ is known and decreases as $O\left(|x|^{-2-\varepsilon}\right)$ at infinity together with the first derivatives, one can find a solenoidal form β which tends to zero at infinity. Such a form is uniquely defined. To this end, we consider the Laplace operator $\Delta=\mathrm{dd}^*+\mathrm{d}^*\mathrm{d}$ and the system of equations

$$\mathrm{d}\beta=\gamma,\ \mathrm{d}^*\beta=0.$$

This system is satisfied if we take the solution of the equation $\Delta\beta=\mathrm{d}^*\gamma$ given by the convolution $F*\mathrm{d}^*\gamma$, where F is the fundamental solution for Δ. The convolution tends to zero at infinity since $F=O\left(|x|^{2-n}\right)$ in the case $n>2$. In the case $n=2$ we need to use the equation $\int\lceil \mathrm{d}^*\gamma\rceil\,\mathrm{d}x=0$. We then have $\mathrm{d}\gamma'=0$, where $\gamma'=\gamma-\mathrm{d}\beta$ and $\Delta\gamma'=\mathrm{dd}^*\gamma-\mathrm{dd}^*\mathrm{d}\beta=\mathrm{d}(\mathrm{d}^*\gamma-\Delta\beta)=0$, which yields $\gamma'=0$. Moreover, $\Delta\mathrm{d}^*\beta=\mathrm{d}^*\mathrm{dd}^*\beta=\mathrm{d}^*\Delta\beta=\mathrm{d}^*\mathrm{d}^*\gamma=0$, since $\mathrm{d}^*\mathrm{d}^*=0$. This yields $\mathrm{d}^*\beta=0$ and completes the proof.

RANGE CONDITION. We say that a function on $\lambda=\lambda(x,\theta)$ defined in the variety $\mathsf{A}_1(\mathbf{V})$ of all affine lines belongs to the Schwartz space, if λ is smooth and any derivative $\mu=D_x^j t_1\dots t_k\lambda$ fulfils the estimate $|\mu|\le C(|x|+1)^{-q}$ for any q as $\langle\theta,x\rangle=0$. Here $t_1,\dots,t_k$ are arbitrary tangent fields on the sphere $\{\theta:|\theta|=1\}$.

Theorem 4.19 **([28])** *A smooth function λ defined on the manifold $\mathsf{A}_1\left(\mathbf{V}^3\right)$ is equal to the line transform of a 1-form α with coefficients in the Schwartz space $S\left(\mathbf{V}^3\right)$, if and only if λ belongs to the Schwartz space and satisfies the equations*

$$\left(\frac{\partial^2}{\partial x_i\partial\theta_j}-\frac{\partial^2}{\partial x_j\partial\theta_i}\right)^2\lambda=0,\ i,j=1,2,3.$$

4.8 Exponential ray transform

Let $\mathbf{E}$ be a Euclidean plane, $\mathbf{S}$ be the unit circle and $a=a(x,\theta)$ be a real function in $\mathbf{E}\times\mathbf{S}$, called an attenuation coefficient in this context. We assume that the ray (Radon) transform

$$Da(x,\theta)=\int_0^\infty a\left(x+t\theta^\perp,\theta\right)\mathrm{d}t,\ x\in\mathbf{E},\,\theta=(\theta_1,\theta_2)\in\mathbf{S},\,\theta^\perp\doteq(-\theta_2,\theta_1)$$

is well defined. The estimate $a(x,\theta) = O\left((|x|+1)^{-\sigma}\right), \sigma > 1$ at infinity is sufficient. Note that we integrate over a ray parallel to the vector $\theta^{\perp}$. For a function f in $\mathbf{E}$ with compact support the integral

$$R_a f(p,\theta) = \int_{\langle\theta,x\rangle=p} f(x)\exp(-Da(x,\theta))\,\mathrm{d}s \tag{4.24}$$

is called the *attenuated* ray transform. The reconstruction problem is to recover f from known functions a and $R_a f$. In this section, we assume that the function a does not depend on $x : a = a(\theta)$ in a disc $\mathbf{B}$ and otherwise vanishes. In this case $Da(x,\theta) = a(\theta)\langle x - x_\theta, \theta^{\perp}\rangle$ where x_θ is the point in the ray $y = x + t\theta^{\perp}, t \geq 0$ where it meets the boundary of $\mathbf{B}$. Therefore

$$R_a f(p,\theta) = \exp(-a(\theta)\langle\theta, x_\theta\rangle)\, E_a f(p,\theta), \tag{4.25}$$

where the first factor does not depend on f and

$$E_a f(p,\theta) \doteq \int_{\langle\theta,x\rangle=p} \exp\left(a(\theta)\langle\theta^{\perp}, x\rangle\right) f(x)\,\mathrm{d}s.$$

The operator E_a is called an *exponential* ray transform. By (4.25) inversion of the attenuated ray transform is reduced to inversion of the operator E_a. Denote $g(x,\theta) = E_a f(\langle\theta,x\rangle,\theta)$.

Theorem 4.20 *If $a \in C^1(\mathbf{S})$, any C^1-function f in a plane with compact support can be reconstructed from g by the formula*

$$f(x) = -\frac{1}{4\pi^2}\int_0^{2\pi} \exp(-a(\theta)\langle\theta,x\rangle)\int_{-\infty}^{\infty} \frac{\cos aq}{q}\left[\frac{\partial}{\partial q} + a'(\theta)\right] g(x + q\theta)\,\mathrm{d}q\,\mathrm{d}\phi,$$

where $a' = \partial a/\partial\phi$ and the inner integral takes its principal value.

Remark. If $a = 0$, the contributions of opposite vectors θ and $-\theta$ are equal and this formula coincides with (2.13).

◀ We write the inversion formula for the plane Fourier transform

$$4\pi^2 f(x) = \int_{\mathbf{E}^*} \hat{f}(\xi)\exp(\imath\langle\xi,x\rangle)\,\mathrm{d}\xi, \tag{4.26}$$

where $\mathbf{E}^*$ is the frequency plane. The integrand is a differential form which admits analytic continuation to the complex space $\mathbb{C}^2 \doteq \mathbf{E}^* + \imath\mathbf{E}^*$ since $\hat{f}$ has analytic continuation in virtue of the Paley-Wiener theorem. Take the chains Π and Δ in $\mathbb{C}^2$ defined as

$$\Pi = (\Gamma,\gamma),\ \Gamma \doteq \{t \in \mathbb{R}, \theta \in \mathbf{S};\, t \geq a(\theta)\},\ \gamma(t,\theta) = \zeta = t\theta + \imath a(\theta)\theta^{\perp}.$$

This is an odd chain (see [18]) with orientation defined by the form $\mathrm{d}t \wedge \mathrm{d}\phi$, where $\theta = (\cos\phi, \sin\phi)$. Define

$$\Delta = (D, \delta), \ D \doteq \{t \in \mathbb{R}, \theta \in \mathbf{S}; 0 \le t \le a(\theta), \text{or } a(\theta) \le t \le 0\},$$
$$\delta(t,\theta) = t\left(\theta + \imath\theta^{\perp}\right).$$

This is also an odd chain oriented by the same form. Boundaries $\partial\Pi$ and $\partial\Delta$ coincide and have opposite orientations. Therefore the sum $\Pi + \Delta$ is a cycle such that we can apply the general integral theorem (see Section 1.6). This allows us to replace the chain $\mathbf{E}^*$ by $\Pi + \Delta$ in (4.26),which yields

$$4\pi^2 f(x) = \int_{\Pi} + \int_{\Delta} \hat{f}(\zeta) \exp(\imath \langle \zeta, x\rangle)\,\mathrm{d}\zeta.$$

Show that the second integral vanishes. We have $\delta(t,\theta) = (\lambda, \imath\lambda)$, where $\lambda = t\theta_1 - \imath t\theta_2$. This implies that the differentials $\mathrm{d}\zeta_1$ and $\mathrm{d}\zeta_2$ are dependent in Δ and the exterior product $\mathrm{d}\zeta = \mathrm{d}\zeta_1 \wedge \mathrm{d}\zeta_2$ vanishes. We have $\zeta = t\theta + \imath a(\theta)\theta^{\perp}$ in the first integral. By Fubini's theorem,

$$\begin{aligned}\hat{f}\left(t\theta + \imath a(\theta)\theta^{\perp}\right) &= \int f(y) \exp\left(a(\theta)\left\langle\theta^{\perp}, y\right\rangle\right) \exp(-\imath t\langle\theta, y\rangle)\,\mathrm{d}y \\ &= \int \exp(-\imath tp) \int_{\langle\theta, y\rangle = p} f(y) \exp\left(a(\theta)\left\langle\theta^{\perp}, y\right\rangle\right) \mathrm{d}s\mathrm{d}p \\ &= \int \exp(-\imath tp)\, g(p\theta, \theta)\,\mathrm{d}p,\end{aligned}$$

which gives an expression of the integral over Π in terms of g. In this way we get

$$4\pi^2 f(x) = \int\int_{a(\theta)}^{\infty} \exp\left(\imath t\langle\theta, x\rangle - a(\theta)\left\langle\theta^{\perp}, x\right\rangle\right) \int \exp(-\imath tp)\,\mathrm{d}t g(y,\theta)\,\mathrm{d}p\mathrm{d}\zeta\mathrm{d}t.$$

A simple calculation gives $\mathrm{d}\zeta = (t + \imath a')\,\mathrm{d}t \wedge \mathrm{d}\phi$ which yields

$$\begin{aligned}4\pi^2 f(x) = &\int \exp\left(-a(\theta)\left\langle\theta^{\perp}, x\right\rangle\right) \\ &\times \int_{-\infty}^{\infty}\int_{a(\theta)}^{\infty} \exp(\imath t(\langle\theta, x\rangle - p))(t + \imath a')\,\mathrm{d}t\, g(p\theta, \theta)\,\mathrm{d}p\mathrm{d}\phi.\end{aligned}$$

Change the variable $q = p - \langle\theta, x\rangle$; we have $p\theta = x + q\theta - \left\langle\theta^{\perp}, x\right\rangle\theta^{\perp}$ and $g(p\theta) = g(x + q\theta)$, since g is constant on any line parallel to $\theta^{\perp}$. Integrating by parts, the inner double integral yields

$$\begin{aligned}\int\int &= \int\int_{a(\theta)}^{\infty} \exp(-\imath tq)(t + \imath a')\,\mathrm{d}t\, g(x + q\theta, \theta)\,\mathrm{d}p \\ &= \int\int \left[\frac{\partial}{\partial q} + a'(\theta)\right] \exp(-\imath tq)\,\mathrm{d}t\, g(x + q\theta, \theta)\,\mathrm{d}p \\ &= -\int \frac{\exp(-\imath a(\theta) q)}{q} \left[\frac{\partial}{\partial q} + a'(\theta)\right] g(x + q\theta, \theta)\,\mathrm{d}q.\end{aligned}$$

This finally gives

$$4\pi^2 f(x) = -\int \exp\left(-a(\theta)\left\langle\theta^\perp, x\right\rangle\right) \times \frac{\exp(-\imath a(\theta) q)}{q}\left[\frac{\partial}{\partial q} + a'(\theta)\right] g(x + q\theta, \theta)\, \mathrm{d}q\mathrm{d}\phi.$$

Taking the real part of the right side, we arrive at (4.26). ▶

The range of the exponential ray transform can be described as follows [1]:

Theorem 4.21 *If $a > 0$ is constant, a function g defined on $\mathbf{E} \times \mathbf{S}$ coincides with the exponential transform of a smooth function f with compact support if and only if g is smooth, has compact support, is constant on each line orthogonal to θ and the Fourier transform $h(\rho, \theta) = F_{p\to\rho} g(p\theta, \theta)$ fulfils the condition $h(\imath\eta, \theta) = h(\imath\eta', \theta')$ whenever $\eta + a\theta = \eta' + a\theta'$.*

4.9 Attenuated ray transform

We now study the reconstruction problem for the attenuated ray transform R_a (see (4.24)) for any real attenuation coefficient a in the plane $\mathbf{E}^2$ that does not depend on the direction. We assume again that the ray transform Da of a is well defined as in the previous section. Define the functions

$$b(p, \theta) = \frac{1}{2}(\mathrm{id} + \imath\mathbf{H}_p) Ra(p, \theta), \quad u(x, \theta) \doteq b(\theta, \langle x, \theta\rangle) - Da(x, \theta),$$

where $\mathbf{H}_p$ is the Hilbert operator applied to a variable p. Note that for arbitrary p, θ and any point x such that $\langle x, \theta\rangle = p$ we have

$$b(p, \theta) + \bar{b}(p, \theta) = Ra(p, \theta) = Da(x, \theta) + Da(x, -\theta). \tag{4.27}$$

The following lemma is due to F. Natterer:

Lemma 4.22 *The function $v(x, \zeta) = u(x, \theta)$, $\zeta = \theta_1 + \imath\theta_2$ has for each x continuation at the disc $\Delta = \{|\zeta| \le 1\}$ that is an odd holomorphic function in the open disc.*

◀ We have

$$u(x, \theta) = \mathrm{j}^{-1}\int_{\mathbf{E}} \frac{a(y)\,\mathrm{d}y}{\langle x - y, \theta\rangle + 0\imath} - \int_0^\infty a\left(x + t\theta^\perp\right) \mathrm{d}t$$
$$= \mathrm{j}^{-1}\left(\int_{\mathbf{E}_+} \frac{a(y)\,\mathrm{d}y}{\langle x - y, \theta\rangle + 0\imath} + \int_{\mathbf{E}_-} \frac{a(y)\,\mathrm{d}y}{\langle x - y, \theta\rangle - 0\imath}\right),$$

where $\mathbf{E}_\pm = \left\{y \in \mathbf{E} : \pm\left\langle x - y, \theta^\perp\right\rangle > 0\right\}$. This follows from the equation for any test function ϕ :

$$\int \frac{\phi(s)\,\mathrm{d}s}{s + 0\imath} - \int \frac{\phi(s)\,\mathrm{d}s}{s - 0\imath} = \mathrm{j}\phi(0).$$

Obviously, we have $u(x,-\theta)=-u(x,\theta)$. Write the integral in terms of the variable ζ and the complex number $z \doteq |x-y|(x_1-y_1+\imath(x_2-y_2))^{-1}$. We have

$$\langle x-y,\theta\rangle = |x-y|\operatorname{Re}\bar{z}\zeta = \frac{|x-y|}{2}\left(\bar{z}\zeta+z\zeta^{-1}\right), \tag{4.28}$$

and

$$v(x,\zeta)=\frac{2}{\mathrm{j}\,|x-y|}\left(\int_{\mathbf{E}_+}\frac{a(y)\,\mathrm{d}y}{\bar{z}\zeta+z\zeta^{-1}+0\imath}+\int_{\mathbf{E}_-}\frac{a(y)\,\mathrm{d}y}{\bar{z}\zeta+z\zeta^{-1}-0\imath}\right). \tag{4.29}$$

The function $\phi(\zeta)\doteq\left(\bar{z}\zeta+z\zeta^{-1}\right)^{-1}=\zeta\left(\bar{z}\zeta^2+z\right)^{-1}$ has holomorphic continuation at the disc Δ. The function $\operatorname{Im}\left(\bar{z}\zeta+z\zeta^{-1}+\varepsilon\imath\right)=\left(|\zeta|^{-2}-1\right)\langle\theta^{\perp},y\rangle$ is strictly positive if $y\in\mathbf{E}_+$ and $\operatorname{Im}\left(\bar{z}\zeta+z\zeta^{-1}-\varepsilon\imath\right)$ is strictly negative if $y\in\mathbf{E}_-$. Therefore both integrals in (4.29) have analytic continuation at Δ. ▶

4.10 Inversion formulae

The following reconstruction result was obtained by R. Novikov:

Theorem 4.23 *A function f in $\mathbf{E}^2$ with compact support can be reconstructed from $g=R_af$ as*

$$f=\frac{1}{4\pi}\operatorname{div}_x\operatorname{Re}R_a^*\left(\exp(b)\,\mathbf{H}_p\exp\left(\bar{b}\right)\tilde{g}\,\theta\right),$$

where $\tilde{g}(p,\theta)=g(-p,-\theta)$ and for a mapping $h=h(p,\theta):\mathsf{A}_1(\mathbf{E})\to\mathbb{C}^2$ we set

$$R_a^*h(x)\doteq\int_{\mathbf{S}^1}\exp(-Da(x,\theta))\,h(\langle x,\theta\rangle,\theta)\,\mathrm{d}\varphi,\ \theta=(\cos\varphi,\sin\varphi). \tag{4.30}$$

◀ By (4.27) we have

$$\begin{aligned}\pi\mathbf{H}_p\exp\left(\bar{b}\right)\tilde{g}(p,\theta)&=\int\frac{\exp\left(\bar{b}(q,\theta)\right)}{p-q}\int_{\langle y,\theta\rangle=q}\exp(-Da(y,-\theta))\,f(y)\,\mathrm{d}s\mathrm{d}q\\&=\int\frac{\exp\left(\bar{b}(\langle y,\theta\rangle,\theta)-Da(y,-\theta)\right)}{p-\langle y,\theta\rangle}f(y)\,\mathrm{d}y\\&=\int\frac{\exp(-b(\langle y,\theta\rangle,\theta)+Da(y,\theta))}{p-\langle y,\theta\rangle}f(y)\,\mathrm{d}y=\int\frac{\exp(-u(y,\theta))}{p-\langle y,\theta\rangle}f(y)\,\mathrm{d}y\end{aligned} \tag{4.31}$$

and

$$\begin{aligned}&\pi R_a^*\left(\theta\exp(b)\,\mathbf{H}_p\exp\left(\bar{b}\right)\tilde{g}\right)(x)\\&=\int_{\mathbf{S}^1}\int\frac{\theta}{\langle x-y,\theta\rangle}\exp(u(x,\theta)-u(y,\theta))\,f(y)\,\mathrm{d}y\mathrm{d}\varphi.\end{aligned} \tag{4.32}$$

Changing the order of integrals we get the interior integral

$$I(x,y)\doteq\int_{\mathbf{S}^1}\frac{\theta}{\langle x-y,\theta\rangle}\exp(u(x,\theta)-u(y,\theta))\,\mathrm{d}\varphi.$$

Introduce the complex variable $\zeta = \theta_1 + \imath\theta_2$; we have $\mathrm{d}\varphi = \mathrm{d}\zeta/\imath\zeta$. By (4.28) we can write the column vector θ as $\left(\zeta^2+1, \imath - \imath\zeta^2\right)^{\mathsf{T}}/2\zeta$. Taking into account that $z\bar{z} = 1$, we get

$$I(x,y) = \frac{z}{\imath\,|x-y|}\int_{\partial\Delta}\begin{pmatrix}\zeta^2+1\\ \imath-\imath\zeta^2\end{pmatrix}\exp\left(v(x,\zeta)-v(y,\zeta)\right)\frac{\mathrm{d}\zeta}{\zeta\left(\zeta^2+z^2\right)}.$$

By Natterer's lemma, the integrand is a meromorphic form in Δ with the poles $\zeta = 0, \pm\imath z$. The principal value integral is equal to the sum of the residue at $\zeta = 0$ and of the mean of two residues at $\zeta = \pm\imath z$ times j:

$$I(x,y) = \frac{2\pi}{|x-y|}\left[\begin{pmatrix}\bar{z}\\ \imath\bar{z}\end{pmatrix} - \frac{1}{2}\begin{pmatrix}\bar{z}-z\\ \imath(\bar{z}+z)\end{pmatrix}\cosh\left(\int_{\complement[x,y]} a\mathrm{d}s\right)\right].$$

Here, we take into account that the function v is odd and the number

$$v(x,\imath z) - v(y,\imath z) = \int_0^\infty a\left(x+t\theta^\perp\right)\mathrm{d}t + \int_0^\infty a\left(y-t\theta^\perp\right)\mathrm{d}t \doteq \int_{\complement[x,y]} a\mathrm{d}s \tag{4.33}$$

is real where $\complement[x,y]$ denotes the complement to the interval $[y,x]$ in the straight line. It follows that

$$\operatorname{Re}\left[\begin{pmatrix}\bar{z}\\ \imath\bar{z}\end{pmatrix} - \frac{1}{2}\begin{pmatrix}\bar{z}-z\\ \imath(\bar{z}+z)\end{pmatrix}\cosh\left(\int_{\complement[x,y]} a\mathrm{d}s\right)\right] = \frac{x-y}{|x-y|},$$

where $x-y$ is written as column and, finally

$$\operatorname{Re} R_a^*\left(\exp(b)\,\mathbf{H}\exp\left(\bar{b}\right)\tilde{g}\,\theta\right) = 2\int\frac{x-y}{|x-y|^2}f(y)\,\mathrm{d}y,$$

$$\operatorname{div}\operatorname{Re} R_a^*\left(\exp(b)\,\mathbf{H}\exp\left(\bar{b}\right)\tilde{g}\,\theta\right) = 4\pi f(x),$$

since

$$\operatorname{div}\frac{x}{|x|^2} = 2\pi\delta_0. \blacktriangleright$$

There is a similar reconstruction formula due to Natterer:

Theorem 4.24 *A function f with compact support can be recovered also as*

$$f = \frac{1}{4\pi}\operatorname{div}_x\operatorname{Re} R_{-a}^*\left(\exp(-b)\,\mathbf{H}_p\exp(b)\,g\,\theta\right), \tag{4.34}$$

where the back projection operator R^* is defined as in (4.30).

◀ We have

$$\begin{aligned}
\pi\mathbf{H}\exp(b)\,g(p,\theta) &= \int\frac{\exp\left(b(q,\theta)\right)}{p-q}\int_{\langle y,\theta\rangle=q}\exp\left(-Da(y,\theta)\right)f(y)\,\mathrm{d}s\mathrm{d}q\\
&= \int\frac{\exp\left(b\left(\langle y,\theta\rangle,\theta\right)-Da(y,\theta)\right)}{p-\langle y,\theta\rangle}f(x)\,\mathrm{d}x\\
&= \int\frac{\exp\left(u(y,\theta)\right)}{p-\langle y,\theta\rangle}f(x)\,\mathrm{d}x
\end{aligned}$$

and

$$\pi R^*_{-a}\left(\theta \exp\left(-b\right) \mathbf{H} \exp\left(b\right) g\right)$$
$$= \int \left(\int_{\mathbf{S}^1} \frac{\theta}{\langle x-y,\theta\rangle} \exp\left(u\left(y,\theta\right) - u\left(x,\theta\right)\right) \mathrm{d}\varphi \right) f\left(y\right) \mathrm{d}y.$$

This expression is similar to (4.32), but the argument of the exponent has opposite sign. The inner integral takes the same value as (4.32). Indeed, the kernel $\theta \langle x-y,\theta\rangle^{-1}$ is the even function of θ and the argument is an odd function because $u\left(x,-\theta\right) = -u\left(x,\theta\right)$. Then we can repeat the arguments of the previous proof. ▶

Remark. It is seen from the proofs that the compact support assumption for the original f can be weakened.

4.11 Range conditions

A necessary condition is a generalization of the moment condition for the Radon transform:

Theorem 4.25 *For a function f in $\mathbf{E}^2$ decreasing rapidly at infinity we have*

$$\int_0^{2\pi} \exp\left(\imath m\phi\right) \left(\int_{-\infty}^{\infty} \exp\left(b\left(p,\theta\right)\right) p^k R_a f\left(p,\theta\right) \mathrm{d}p \right) \mathrm{d}\phi = 0 \tag{4.35}$$

for any natural $m > k \geq 0$.

Remark. The equation holds also if we replace m by $-m$ and b by $\bar{b}$ since $R_a f$ is a real function. If $a = 0$ these conditions coincide with Proposition 2.10.

◀ We can write

$$\exp\left(b\left(p,\theta\right)\right) R_a f\left(p,\theta\right) = \int_{\langle x,\theta\rangle = p} \exp\left(u\left(x,\theta\right)\right) f\left(x\right) \mathrm{d}x,$$

$$\int_{-\infty}^{\infty} p^k \exp\left(b\left(p,\theta\right)\right) R_a f\left(p,\theta\right) \mathrm{d}p = \int \langle x,\theta\rangle^k \exp\left(u\left(x,\theta\right)\right) f\left(x\right) \mathrm{d}x$$
$$= 2^{-k} \int \left(\bar{w}\zeta + w\zeta^{-1}\right)^k \exp\left(v\left(x,\zeta\right)\right) f\left(x\right) \mathrm{d}x,$$

where $w \doteq x_1 + \imath x_2$. The kernel $\left(\bar{w}\zeta + w\zeta^{-1}\right)^k \exp\left(v\left(x,\zeta\right)\right)$ has analytic continuation at the disc $\Delta \setminus \{0\}$ as a function of ζ. The continuation has a pole at $\zeta = 0$ of order at most k. Therefore the integral

$$\int_0^{2\pi} \left(\bar{w}\zeta + w\zeta^{-1}\right)^k \exp\left(v\left(x,\zeta\right)\right) \exp\left(\imath m\phi\right) \mathrm{d}\phi$$
$$= -\imath \int_{\partial\Delta} \left(\bar{w}\zeta + w\zeta^{-1}\right)^k \exp\left(v\left(x,\zeta\right)\right) \zeta^{k-1} \mathrm{d}\zeta$$

vanishes for $m > k$. This implies (4.35). ▶

The following range condition is written in a form similar to the reconstruction formula.

Theorem 4.26 *For an arbitrary continuous function f with compact support in $\mathbf{E}^2$, the following equation holds for all $x \in \mathbf{E}$:*

$$\operatorname{Re} \int_0^{2\pi} \exp(-Da(x,\theta)) \exp(b(\langle x,\theta\rangle,\theta)) \mathbf{H}_{q \mapsto \langle x,\theta\rangle} \exp\left(\bar{b}(q,\theta)\right) g(-q,-\theta) \mathrm{d}\varphi = 0, \tag{4.36}$$

where $g = R_a f$.

◀ Denote by J the left side. By (4.32) we have

$$J = \operatorname{Re} \int \left(\int_{\mathbf{S}^1} \frac{\mathrm{d}\varphi}{\langle x-y,\theta\rangle} \exp\left(u(x,\theta) - u(y,\theta)\right) \right) f(y)\, \mathrm{d}y.$$

Calculate the interior integral by means of the complex variable $\zeta = \theta_1 + \imath\theta_2$, taking into account that $\mathrm{d}\varphi = \mathrm{d}\zeta / \imath\zeta$:

$$\int_{\mathbf{S}^1} \frac{\mathrm{d}\varphi}{\langle x-y,\theta\rangle} \exp(u(x,\theta) - u(y,\theta)) = \frac{z}{\imath|x-y|} \int_{\partial\Delta} \exp(v(x,\zeta) - v(y,\zeta)) \frac{\mathrm{d}\zeta}{\zeta^2 + z^2}.$$

This principal value integral is equal to the mean of the residues at $\zeta = \pm\imath z$ times j which gives the number

$$\frac{\pi}{2\imath|x-y|} \left[\exp\left(v(x,\imath z) - v(y,\imath z)\right) - \exp\left(v(x,-\imath z) - v(y,-\imath z)\right)\right]$$

$$= \frac{\pi}{2\imath|x-y|} \sinh\left(\int_{\mathsf{C}[x,y]} a \mathrm{d}s \right),$$

where the quantity

$$v(x,\imath z) - v(y,\imath z) = v(x,\imath z) + v(y,-\imath z) = \int_{\mathsf{C}[x,y]} a \mathrm{d}s$$

is real. Therefore the right side is pure imaginary and the real part of the left side vanishes. ▶

A similar range condition can be written in terms of Natterer's formula:

Theorem 4.27 *Under the same conditions the equation holds for all $x \in \mathbf{E}^2$,*

$$\operatorname{Re} \int_0^{2\pi} \exp(Da(x,\theta)) \exp(-b(\langle x,\theta\rangle,\theta)) \mathbf{H}_{q \mapsto \langle x,\theta\rangle} \exp(b(q,\theta)) g(-q,-\theta) \mathrm{d}\varphi = 0. \tag{4.37}$$

◀ A proof can be given along the same lines as in the previous theorem. ▶

Remark 1. The equations (4.36) and (4.37) can be extended for arbitrary functions f in $\mathbf{E}$ such that the integrals in the left sides converge absolutely. These equations turn to the evenness condition $g(-p,-\theta) = g(p,\theta)$ as a vanishes.

Remark 2. It is not known whether the set of the range conditions given in the three last theorems are complete. The problem can be formulated as follows: let $g = g(p,\theta)$ be an arbitrary continuous function in $\mathbb{R}\times\mathbf{S}^1$ that fulfils (4.35), (4.36) and (4.37) and vanishes for $|p| > r$ for some r. Is it true that there always exists a function f in $\mathbf{E}$ supported by a ball of radius r such that $g = R_a f$?

On the other hand, if the attenuation coefficient is constant, the complete description of the range of the exponential ray transform in $\mathbb{R}^n$ is known, see [1].

Chapter 5

Flat Integral Transform

5.1 Reconstruction problem

Let $\mathbf{E}$ be a Euclidean space of dimension n. Consider the manifold $\mathsf{A}(\mathbf{E})$ of all affine subspaces $A \subset \mathbf{E}$. Let $\mathrm{d}V(A)$ be the Euclidean volume density in A. The operator

$$f \mapsto Mf(A) \doteq \int_A f \, \mathrm{d}V(A), \quad A \in \mathsf{A}(\mathbf{E})$$

defined for functions f that decrease sufficiently fast at infinity is called a *flat* integral transform in $\mathbf{E}$. Fix a natural $k < n$ and consider the submanifold $\mathsf{A}_k(\mathbf{E})$ of affine subspaces of dimension k. It is an algebraic variety of dimension $(k+1)(n-k)$. Denote by M_k the restriction of M to $\mathsf{A}_k(\mathbf{E})$. If $k = n-1$ we keep the notation of the Radon transform $Rf = M_{n-1}f$. We have discussed the reconstruction problem for the transform M_{n-1} in Chapter 2 and for the operator M_1 in Chapter 4. The inversion problem for the operator $M_k f$, $k < n-1$ immediately reduces to the case $k = n-1$. Indeed, we reconstruct $M_{k+1}f$ from $M_k f$ by inversion of the Radon transform in each $k+1$-plane in $\mathbf{E}$. On the other hand, the scope of integrals $M_k f$ is redundant for reconstruction of f if $k < n-1$, since $\dim \mathsf{A}_k(\mathbf{E}) = (k+1)(n-k) > n$. Therefore there is a large variety of inversion methods for $M_k f$. To avoid redundancy we state the reconstruction problem as follows:

Problem: to find a reconstruction formula $Mf \mid \Sigma \mapsto f$ for functions f with compact support in $\mathbf{E}$ and a submanifold $\Sigma \subset \mathsf{A}(\mathbf{E})$ of dimension n (called a *pencil*). The set of integrals $Mf(L)$, $L \in \Sigma$ has no dimension redundancy.

Let now $\mathbf{V}$ be an arbitrary real vector space of dimension n. For a natural $k < n$, let $\mathbf{Gr}_k^*(\mathbf{V})$ denote the manifold of all oriented k-dimensional subspaces L in $\mathbf{V}$ and $\mathbf{Gr}_k(\mathbf{V})$ be the manifold of non-oriented k-subspaces (Grassmann manifolds). There is defined the double covering $\mathbf{p} : \mathbf{Gr}_k^* \to \mathbf{Gr}_k$. There is the natural projection $G_k : \mathsf{A}_k(\mathbf{V}) \to \mathbf{Gr}_k$ which brings a k-plane to the parallel subspace. A k-frame in $\mathbf{V}$ is an ordered set of k linearly independent vectors $\theta = (\theta_1, \ldots, \theta_k)$. Let $\mathbf{Fr}_k$ be the manifold of all k-frames in $\mathbf{V}$; there is a natural mapping $\mathbf{q} : \mathbf{Fr}_k \to \mathbf{Gr}_k^*$ that sends a frame to its linear envelope oriented by the vectors $\theta_1, \dots, \theta_k$. Introduce the differential operator in $\mathbf{Fr}_k \times \mathbf{V}$:

$$K(\theta) = \langle \mathrm{d}\theta_1, \partial_x \rangle \wedge \cdots \wedge \langle \mathrm{d}\theta_k, \partial_x \rangle, \ \partial_x = (\partial/\partial x_1, \dots, \partial/\partial x_n).$$

Let f be a continuous function in $\mathbf{V}$ that decreases sufficiently fast at infinity. For any point $x \in \mathbf{V}$ and any k-frame θ we can define the integral

$$If(x, \theta) \doteq \int_\theta f(x + t_1\theta_1 + \cdots + t_k\theta_k) \, \mathrm{d}t_1 \dots \mathrm{d}t_k.$$

Obviously, we have $If(x, \eta) = |\det A|^{-1} If(y, \theta)$ where $\eta = A\theta$ is another k-frame in the subspace L generated by θ, and $y - x \in L$.

Suppose that $\mathbf{V}$ is endowed with a structure of Euclidean space. A finite set $\theta = \{\theta_i\} \subset \mathbf{V}$ fulfilling the condition $\langle \theta_i, \theta_j \rangle = \delta_{ij}$ is called an *orthoframe*. For a function f in $\mathbf{E}$ we have $Mf(P) = If(x, \theta)$ where $P = x + L$ and θ is an orthoframe in the subspace L. Therefore the reconstruction problem can be reformulated and generalized in the terms of the integral transform I for an arbitrary real vector space $\mathbf{V}$.

Let θ, η be orthoframes in L that define the same orientation in the space L. We have $\eta = \Lambda\theta$ where Λ is a $k \times k$-matrix such that $\det \Lambda = 1$. Obviously $K(\eta) = \det A \ K(\theta) = K(\theta)$.Therefore the operator K is well defined also on the manifold $\mathsf{A}_k^*(\mathbf{V})$ of oriented k-planes $P \subset \mathbf{E}$.

5.2 Odd-dimensional subspaces

Let $\mathbf{E}$ be again a Euclidean space of dimension n. Consider the differential form in $\mathbf{Fr}_k \times \mathbf{E}^*$,

$$\Delta(\theta) = \Delta(\theta, \xi) \doteq \langle \mathrm{d}\theta_1, \xi \rangle \wedge \cdots \wedge \langle \mathrm{d}\theta_k, \xi \rangle,$$

where $\theta = (\theta_1, \dots, \theta_k)$ is a k-frame in $\mathbf{E}$. The exterior product is an even k-form on $\mathbf{Fr}_k$, whose coefficients are homogeneous polynomials of degree k. It pulls down to $\mathbf{Gr}_k \times \mathbf{E}^*$: for each oriented subspace $L \subset \mathbf{E}$ we can choose a frame θ in L that is a smooth (rational) function of L. Changing the frame $\eta = A\theta$ by means of a linear transformation A makes the transformation $\Delta(\eta) = \det A \ \Delta(\theta)$, where $\det A = 1$. This implies that $\Delta(\eta) = \Delta(\theta)$ and we may denote the form by $\Delta(L)$. Note that for any L the coefficients of this form depend only on the projection η of ξ to $L^\perp$. Indeed, if $\xi = \eta + \zeta, \zeta \in L$, then $\langle \mathrm{d}\theta_j, \zeta \rangle = 0$. For any k-frame of tangent vectors $t_1, \dots, t_k$ to $\mathbf{Gr}_k^*$ at a point L, we consider the functional

$$|\Delta(L)|(t_1, \dots, t_k) = |\Delta(L)(t_1, \dots, t_k)|$$

depending on k tangent vectors $t_1, \dots, t_k$ in $\mathbf{Fr}_k$. It is positively homogeneous in each argument and symmetric with respect to permutations. Define the singular kernel in $L^\perp$:

$$D(L, r)(t_1, \dots, t_k) \doteq \int_{L^\perp} |\Delta(L, \eta)(t_1, \dots, t_k)| \exp(-\mathrm{j}\langle \eta, r \rangle) \mathrm{d}\eta, \tag{5.1}$$

where $\mathrm{d}\eta$ is the Euclidean volume form in $L^{\perp}$. Thus $D(L,r)$ is an odd form on each k-dimensional manifold in $\mathbf{Fr}_k$ with values in the space of distributions in $\mathbf{E}$.

Take a submanifold $\mathbf{C} \subset \mathbf{Gr}_k$ of dimension k and consider the variety $\Sigma(\mathbf{C}) \doteq G_k^{-1}(\mathbf{C})$. It consists of k-planes of the form $P = x + L$, where $L \in \mathbf{C}$, $x \in L^{\perp}$ ($L^{\perp}$ means the orthogonal complement to L). Therefore $\dim \Sigma(\mathbf{C}) = n$; data of integrals $Mf(P), P \in \Sigma(\mathbf{C})$ are non-redundant.

Take a subspace $\mathbf{F}$ in $\mathbf{V}$ of dimension $k+1$; the manifold $\mathbf{Gr}_k^*(\mathbf{F})$ of oriented k-subspaces $L \subset \mathbf{F}$ has dimension $k+1$. It is isomorphic to the sphere $\mathbf{S}^k$. Denote by $\mathbf{P}_k$ the corresponding homotopy class in $\mathbf{Gr}_k^*(\mathbf{E})$.

Theorem 5.1 *Suppose that k is odd and $\mathbf{C}$ is a k-cycle in $\mathbf{Gr}_k(\mathbf{E})$ that belongs to the class homotopy of $\mathbf{P}_k$. The formula*

$$f(x) = \int_{\mathbf{C}} \int_{L^{\perp}} D(L,r)\; g(r + x + L)\mathrm{d}r \tag{5.2}$$

gives reconstruction for any function $f \in C^{k+1}(\mathbf{E})$ with compact support from data of its affine transform $g(P) = Mf(P)$, $P \in \Sigma(\mathbf{C})$.

Remark. The function $g(\cdot + L + x)$ belongs to C^{k+1} and the kernel $D(L,\cdot)$ can be applied.

◀ Take the cycle $\mathbf{C}^* \doteq \mathbf{p}^{-1}(\mathbf{C}) \subset \mathbf{Gr}_k^*$ and decompose it into two chains $\mathbf{C}^* = \mathbf{C}_+ \cup \mathbf{C}_-$ in such a way that $\mathbf{p} : \mathbf{C}_{\pm} \to \mathbf{C}$ are bijections except for a zero measure subset. For this we choose an $n-k$-subspace $\mathbf{G} \subset \mathbf{E}$ and take the cell $\mathbf{G}^*$ in $\mathbf{Gr}_k^*$ consisting of k-planes that are not transversal to $\mathbf{G}$. The open set $\mathbf{Gr}_k^* \backslash \mathbf{G}^*$ consists of two connected components such that $\mathbf{p}$ is injective on each of them. The intersection $\mathbf{C} \cap \mathbf{G}$ has zero measure in $\mathbf{C}$, if $\mathbf{G}$ is chosen properly and we take for $\mathbf{C}_{\pm}$ the intersection of $\mathbf{C} \cap \mathbf{G}$ with the leaves. Choose an orientation in the subspace $\mathbf{F} \subset \mathbf{E}$; it induces an orientation in the $\mathbf{Gr}_k^*(\mathbf{F})$ and in $\mathbf{C}^*$by means of the homotopic equivalence $\mathbf{C}^* \approx \mathbf{Gr}_k^*(\mathbf{F})$. Let σ be an orientation form in $\mathbf{C}^*$ which is preserved by the natural bijection $\mathbf{b} : \mathbf{C}_+ \to \mathbf{C}_-$. Choose an orthoframe $\theta = (\theta_1, \ldots, \theta_k)$ in L depending smoothly on $L \in \mathbf{C}^*$. Choose a Euclidean coordinate system $\xi = (\xi_1, \ldots, \xi_n)$ in the dual space $\mathbf{E}^*$ and take $\upsilon \doteq \mathrm{d}\xi_1 \wedge \cdots \wedge \mathrm{d}\xi_n$ as the orientation form in $\mathbf{E}^*$. Consider the $n+k$-manifold $\mathbf{E}^* \times \mathbf{C}^*$ oriented by the form $\upsilon \wedge \sigma$. The restriction of the form $\mathrm{d}\langle\theta_i, \xi\rangle$ to $\mathbf{E}^* \times \mathbf{C}^*$ is equal to $\langle\theta_i, \mathrm{d}\xi\rangle + \langle \mathrm{d}\theta_i, \xi\rangle$. The forms $\langle\theta_1, \mathrm{d}\xi\rangle, \ldots, \langle\theta_k, \mathrm{d}\xi\rangle$ are independent in $\mathbf{E}^*$ for each L and therefore the form $K = \langle\theta_1, \mathrm{d}\xi\rangle \wedge \cdots \wedge \langle\theta_k, \mathrm{d}\xi\rangle$ does not vanish. We can find a global $n-k$-form ψ in $\mathbf{E}^* \times \mathbf{C}^*$ that satisfies $\psi \wedge K = \upsilon$, since this condition does not depend on the choice of the frame θ. The restriction of ψ to the subspace L generated by the frame θ is well defined and coincides up to sign with the Euclidean volume density in L. Consider the manifold $\Gamma \doteq \left\{(L,\xi) : L \in \mathbf{C}^*, \xi \in L^{\perp}\right\} \subset \mathbf{E}^* \times \mathbf{C}^*$ and the projection $\gamma_{\mathbf{C}} : \Gamma \to \mathbf{E}^*$, $\gamma_{\mathbf{C}}(L,\xi) = \xi$. We claim that $\gamma_{\mathbf{C}}^*(\upsilon) = \omega$ in Γ, where $\omega = -\psi \wedge \Delta$. Indeed, the forms $\rho \doteq \upsilon - \omega$ and $\Theta \doteq \mathrm{d}\langle\theta_1, \xi\rangle \wedge \cdots \wedge \mathrm{d}\langle\theta_k, \xi\rangle$ are defined in $\mathbf{E}^* \times \mathbf{C}^*$ and fulfil the equation

$$\rho \wedge \Theta = \upsilon \wedge \Delta + \psi \wedge \Delta \wedge K = (\upsilon - \psi \wedge K) \wedge \Delta = 0,$$

since $\Delta \wedge K = -K \wedge \Delta$. This implies that $\rho = \sum \mathrm{d}\langle \theta_j, \xi \rangle \wedge \rho_j$ for some forms $\rho_1, \ldots, \rho_k$. The forms $\mathrm{d}\langle \theta_1, \xi \rangle, \ldots, \mathrm{d}\langle \theta_1, \xi \rangle$ vanish on Γ. Therefore $\rho = 0$ and $\gamma_{\mathbf{C}}^* (v) = \omega$ in Γ. Define the function $\delta \doteq \Delta/\sigma$ in $\mathbf{C}_* \times \mathbf{E}^*$ and consider the domain in Γ

$$\Gamma_+ \doteq \{(L, \xi) \in \Gamma : \ \delta(L, \xi) > 0\}$$

oriented by the form ω, and consider the mapping $\gamma_{\mathbf{C}} : \Gamma_+ \to \mathbf{E}^*$, $\gamma_{\mathbf{C}}(L, \xi) = \xi$. The pair $\Pi = (\Gamma_+, \gamma_{\mathbf{C}})$ is an odd n-chain in $\mathbf{E}^*$.

Lemma 5.2 *The chain* Π *is closed.*

◀ The boundary $\partial \Gamma_+$ in Γ is given by the equation $\delta(L, \xi) = 0$; the form $\mathrm{d}\delta(L, \xi)$ vanishes for all $(L, \xi) \in \mathbf{C}^* \times \mathbf{E}^*$, except for a subset of zero measure, which implies that $\partial \Gamma_+$ is smooth almost everywhere. The orientation of the boundary is defined by a form χ such that $-\mathrm{d}\delta \wedge \chi = \omega = -\delta\psi \wedge \sigma$. Let $L^- \in \mathbf{C}_-$ be the point opposite to a point $L \in \mathbf{C}_+$, i.e., $\mathbf{p}(L^-) = \mathbf{p}(L)$. The orientation in L^- is opposite to that of L and we can take $-\theta$ as the orthoframe in L^-. Therefore $\delta(L^-, \xi) = -\delta(L, \xi)$ and the domain $\Gamma_+(L^-) \doteq \left\{\xi \in L^\perp, \delta(L^-, \xi) > 0\right\}$ is complementary to $\Gamma_+(L)$. The orientation of $\partial \Gamma_+$ near L^- is given by the form χ^- that fulfils the similar equation. We have $\psi(L^-) = -\psi(L)$, which implies that $\chi^- = -\chi$ and the orientation is opposite. ▶

The mapping $\gamma_{\mathbf{C}}$ is proper and the topological degree $\deg \gamma_{\mathbf{C}}$ is well defined.

Lemma 5.3 $\deg \gamma_{\mathbf{C}} = 1$.

◀ Denote this mapping by $\gamma_{\mathbf{C}}$. The degree of $\gamma_{\mathbf{C}}$ changes if we replace $\mathbf{C}$ by any homologically equivalent cycle. It follows that $\deg \gamma_{\mathbf{C}} = \deg \gamma_{\mathbf{P}_k}$. The mapping $\gamma_{\mathbf{P}_k}$ restricted to the interior of Γ is a bijection onto $\mathbf{E}^* \backslash \mathbf{F}^\perp$. The form $\omega' = v/\delta(L, \xi)$ fulfils the equation $\omega' \wedge \Delta = v \wedge \sigma$, which implies that ω defines the orientation of Γ_+. It coincides with the orientation of $\mathbf{E}$ since $\delta(L, \xi) > 0$ in Γ_+. This yields the equation $\deg \gamma_{\mathbf{P}_k} = 1$. ▶

Write the Fourier transform of an unknown function

$$\hat{f}(\xi) = \int_{\mathbf{E}} \exp(-\mathrm{j}\langle \xi, x \rangle) f(x) \mathrm{d}x,$$

where $\mathrm{d}x$ is the Euclidean volume density. Take a subspace L orthogonal to ξ, and an orthoframe θ in L. Write $x \in y + L, y \in L^\perp$ and integrate over L and $L^\perp$ consecutively, taking into account that $\langle \xi, x \rangle = \langle \xi, y \rangle$:

$$\hat{f}(\xi) = \int_{L^\perp} \exp(-\mathrm{j}\langle \xi, y \rangle) g(y + L) \mathrm{d}r, \tag{5.3}$$

where $\mathrm{d}r$ is the Euclidean area element in $L^\perp$. Lemma 5.3 implies that

$$f(x) = \int_{\mathbf{E}} \exp(\mathrm{j}\langle \xi, x \rangle) \hat{f}(\xi)\, v = \int_{\Gamma_+} \exp(\mathrm{j}\langle \xi, x \rangle) \hat{f}(\xi) \gamma^* (v)\,. \tag{5.4}$$

We replace $\gamma^*(v)$ by $\omega = -\psi \wedge \Delta$ in (5.4) and substitute (5.3):

$$\int_{\Gamma_+} \exp(\mathrm{j}\langle\xi, x\rangle) \hat{f}(\xi) \gamma^*(v)$$
$$= -\int_{\mathbf{C}^*} \sigma \int_{\Gamma_+(L)} \delta(L,\xi) \exp(\mathrm{j}\langle\xi, x-y\rangle) \psi \int_{L^\perp} g(y+L)\mathrm{d}r.$$

Because of the equation $\omega = -\delta\psi \wedge \sigma$, the form $-\psi$ defines the orientation of the $n-k$-domain $\Gamma_+(L)$ for any $L \in \mathbf{C}^*$. We can write $y - x = q + r$, where $q \in L$, $r \in L^\perp$, which yields $\langle\xi, x-y\rangle = \langle\eta, r\rangle$, where η is the orthogonal projection of ξ to $L^\perp$. We have $g(y+L) = g(r+x+L)$ and $\delta(L,\xi) = \delta(L,\eta)$. This yields

$$f(x) = \int_{\mathbf{C}^*} \sigma \int_{L^\perp} \int_{\Gamma_+(L)} \delta(L,\eta) \exp(\mathrm{j}\langle\eta, r\rangle)\, g(r+x+L)\mathrm{d}r\, \mathrm{d}\eta,$$

where $\mathrm{d}\eta$ is the Euclidean volume density in $L^\perp$. Now we take the average over each pair of opposite planes $L \in \mathbf{C}_+$ and $L^- \in \mathbf{C}_-$. We have

$$\theta(L^-) = -\theta(L),\ \delta\left(L^-, \xi\right) = -\delta(L,\xi),\ \Gamma_+\left(L^-\right) = \Gamma_-(L) \doteq \{\xi : \delta(L,\xi) < 0\}.$$

This yields

$$f(x) = \int_{\mathbf{C}_+} \sigma \left(\int_{\Gamma_+(L)} - \int_{\Gamma_-(L)}\right) \delta(L,\eta) \exp(-\mathrm{j}\langle\eta, r\rangle) \int_{L^\perp} g(r+x+L)\mathrm{d}r\, \mathrm{d}\eta$$
$$= \int_{\mathbb{C}_+} \sigma \int_{L^\perp} \int_{L^\perp} |\delta(L,r)| \exp(-\mathrm{j}\langle\eta, r\rangle)\, \mathrm{d}\eta\, g(r+x+l)\mathrm{d}r. \tag{5.5}$$

The sum is equal to

$$f(x) = \int_{\mathbf{C}_+} \int |\Delta(L,\eta)| \exp(-\mathrm{j}\langle\eta, r\rangle) \int_{L^\perp} g(r+x+L)\mathrm{d}r\ \mathrm{d}\eta.$$

We replace $\mathbf{C}_+$ by the isomorphic chain $\mathbf{C}^*$ and get (5.2). ▶

Example 1. Take a k-dimensional subspace $\mathbf{F} \subset \mathbf{E}$ and the chain $\mathbf{C} = \mathbf{Gr}_k(\mathbf{F})$. The function $g(P), P \in \Sigma(\mathbf{P}_k)$ is the Radon transform of f in each space $\mathbf{F} + z$. Then $\Delta(L,\xi) = |\langle\nu(L), \xi\rangle|^k \Omega$ where $\nu(L)$ is the unit normal vector to L and Ω is the volume density in the unit sphere in $\mathbf{F}^*$. Therefore

$$\Delta(L,r) = \frac{(-1)^{(k-1)/2}}{(2\pi)^k} \mathbf{H} \left(\frac{\partial}{\partial p}\right)^k \Omega,$$

where $\mathbf{H}$ is the Hilbert operator in the p-direction. Then (5.2) coincides with (2.6) where $\mathbf{S}^k = \mathbf{C}^*$.

Example 2. Let $k = 1$ and take for $\mathbf{C}$ the image in $\mathbf{Gr}_1(\mathbf{E})$ of a curve $\mathbf{C}_a \subset \mathbf{S}^{n-1}$ that joins opposite points a and $-a$. Then

$$D(L, r) = \int_{\theta^\perp} |\langle \mathrm{d}\theta, \eta \rangle| \exp(-\mathrm{j}\langle \eta, r \rangle)\, \mathrm{d}r = -\frac{1}{2\pi} \mathbf{H}_{\mathrm{d}\theta} \left\langle \mathrm{d}\theta, \frac{\partial}{\partial r} \right\rangle.$$

Choose a parameterization $\theta = \theta(s)$ of $\mathbf{C}_a$; then $\mathrm{d}\theta(\partial/\partial s) = \theta'(s)$ and $\mathbf{H}_{\mathrm{d}\theta}$ is the Hilbert transform in the direction $\theta'(s)$. The formula (5.2) yields

$$f(x) = -\frac{1}{2\pi^2} \int_{\mathbf{C}_a} \int \frac{\partial}{\partial q} \int_{\mathbb{R}} g(x + q\theta'(s) + L) \frac{\mathrm{d}q}{q},$$

which agrees with Theorem 4.3.

5.3 Even dimension

Theorem 5.4 *If k is even and $\mathbf{C}$ is a cycle in $\mathbf{Gr}_k(\mathbf{E})$ that belongs to the homology class of $\mathbf{P}_k$, then the formula*

$$f(x) = \mathrm{j}^k \int_{\mathbf{C}} K\left(L, \frac{\partial}{\partial x}\right) g(x + L) \tag{5.6}$$

gives reconstruction for any function $f \in C^k(\mathbf{E})$ with compact support from data of integrals $g(P) = Mf(P)$, $P \in \Sigma(\mathbf{C})$.

◀ Keeping the notation of the previous arguments, consider the chain

$$\begin{aligned} \Gamma^* &= \left\{(L, \xi) : L \in \mathbf{C}_+, \xi \in L^\perp, \delta(L, \xi) > 0\right\} \\ &\cup \left\{(L, \xi) : L \in \mathbf{C}_-, \xi \in L^\perp, \delta(L, \xi) < 0\right\}. \end{aligned}$$

We have now $\Delta(L^-) = \Delta(L)$, $\psi(L^-) = \psi(L)$ and define $\omega = \psi \wedge \Delta$. The arguments of Lemma 5.2 show that the chain Γ^* is closed and by arguments of Lemma 4.5, $\deg \gamma_{\mathbf{C}} = 1$ where $\gamma_C(L, \xi) = \xi$ for $(L, \xi) \in \Gamma^*$. Then we repeat the calculations which yield instead of (5.5) the equation

$$\begin{aligned} f(x) &= \int_{\mathbf{C}_+} \int_L \Delta(\theta, \xi) \exp(-\mathrm{j}\langle \xi, r \rangle)\, \eta \int_{L^\perp} g(r + L)\mathrm{d}r \\ &= \int_{\mathbb{C}} K\, g(x + L), \end{aligned}$$

since Θ is the symbol of the homogeneous operator K. ▶

Remark. The operator K is known under the name "operator κ", see [29]. The reconstruction (5.6) was obtained as a corollary of the following key property of this operator:

Proposition 5.5 *The restriction of Kg to $\mathbf{Gr}_k(x)$ is a closed form for any point $x \in \mathbf{E}$.*

◀ Take a k-plane $P = x + L$ and choose an orthoframe $\theta = (\theta_1, \ldots, \theta_k)$ in the subspace L. We have

$$Kg(P) = \langle \mathrm{d}\theta_1, \partial_x \rangle \wedge \cdots \wedge \langle \mathrm{d}\theta_k, \partial_x \rangle \int_{\mathbf{R}^k} f(x + t_1\theta_1 + \cdots + t_k\theta_k)\, \mathrm{d}t,$$

where $\mathrm{d}t = \mathrm{d}t_1 \ldots \mathrm{d}t_k$ and

$$\begin{aligned} \mathrm{d}_\theta Kg(P) &= (-1)^k \int_{\mathbf{R}^k} \langle \mathrm{d}\theta_1, \partial_x \rangle \wedge \cdots \wedge \langle \mathrm{d}\theta_k, \partial_x \rangle \wedge \mathrm{d}_\theta f(x + t_1\theta_1 + \cdots + t_k\theta_k)\, \mathrm{d}t \\ &= (-1)^k \int_{\mathbf{R}^k} \langle \mathrm{d}\theta_1, \partial_x \rangle \wedge \cdots \wedge \langle \mathrm{d}\theta_k, \partial_x \rangle \\ &\wedge \sum_j \langle \mathrm{d}\theta_j, \partial_x \rangle t_j f(x + t_1\theta_1 + \cdots + t_k\theta_k)\, \mathrm{d}t. \end{aligned}$$

The right side vanishes, since $\langle \mathrm{d}\theta_j, \partial_x \rangle \wedge \langle \mathrm{d}\theta_j, \partial_x \rangle = 0$ for any j, which means that $\mathrm{d}_\theta Kg = 0$. ▶

5.4 Range of the flat transform

A description of the range of the operator M_k can be done in two ways:

MOMENT CONDITIONS. Take the following parameterization of the manifold $\mathsf{A}_d = \mathsf{A}_d(\mathbf{E})$ of all affine subspaces A of $\mathbf{E}$. Take the vector subspace A_0 of $\mathbf{E}$ of dimension d that is parallel to A; denote by $A^\perp$ the orthogonal complement to A_0. The intersection $A \cap A^\perp$ is always one point y; the pair (A_0, y) is the parameterization of A_d. Here A_0 runs over the manifold $\mathbf{Gr}_d(\mathbf{E})$ of d-subspaces of $\mathbf{E}$ and y runs in $A^\perp = A_0^\perp$.

Let g be a function defined on the manifold A_d. We say that g is fast decreasing at infinity, if $g(A) = O\left(|y|^{-q}\right)$ for arbitrary $q > 0$; note that $|y|$ is equal to the distance between A and the origin. The function fulfils the moment conditions if it decreases fast at infinity and for an arbitrary integer $m \geq 0$ there exists a homogeneous polynomial P_m on $\mathbf{E}$ of degree m such that for any $A \in \mathsf{A}_d$ it holds that

$$P_m(\eta) = \int_{A^\perp} g(y, A_0) \langle y, \eta \rangle^m \,\mathrm{d}y. \tag{5.7}$$

In other words, the right-hand side does not depend on A_0.

Theorem 5.6 (Helgason) *A function g defined on A_d is equal to $M_d f$ for a function f in $\mathbf{E}$ with compact support if and only if it has compact support and satisfies the moment conditions.*

The main point is the sufficiency of (5.7).

JOHN'S EQUATIONS. F. John has proved that the system of equations (4.2) for a function g in $\mathsf{A}_1(\mathbf{E}^3)$ that belongs to the Schwartz space implies that $g = M_1 f$ of a function $f \in S(\mathbf{E}^3)$, see [47].

GENERALIZED JOHN EQUATIONS. The generalization of John's equations is a system of second-order equation of a similar structure. To write it explicitly, we introduce the parameterization of the manifold $\mathsf{A}_d(\mathbf{E})$ of all affine subspaces in $\mathbf{E}$ of dimension d. Take a subspace $A \in \mathsf{A}_d(\mathbf{E})$ and choose $d+1$ points $y_0, \ldots, y_d \in A$ in general position. The space A is generated by these points in the sense that $A \doteq \{x : x = \sum t_j y_j, \sum t_j = 1\}$. In other words, A is the shift by y_0 of the vector subspace in $\mathbf{E}$ generated by the vectors $v_1 \doteq y_1 - y_0, \ldots, v_d \doteq y_d - y_0$. Let $\mathrm{Vol}(v_1, \ldots, v_d)$ be the Euclidean volume of the frame: $\mathrm{Vol}^2(v_1, \ldots, v_d) = \sum m^2$, where the sum is taken over the set of d-minors m of the $d \times n$ matrix $(v_1, \ldots, v_d)$. We write $M_k f(y_0, \ldots, y_d)$ for the integral $M_k f(A)$.

Proposition 5.7 *For any rapidly decreasing function f in $\mathbf{E}$ the following equations hold:*

$$\left(\frac{\partial}{\partial y_{i,k}} \frac{\partial}{\partial y_{j,l}} - \frac{\partial}{\partial y_{i,l}} \frac{\partial}{\partial y_{j,k}}\right) \frac{M_k f(y_0, \ldots, y_d)}{\mathrm{Vol}(v_1, \ldots, v_k)} = 0, \; i, j = 0, \ldots, d, \; k, l = 1, \ldots, n. \tag{5.8}$$

A proof is similar to the arguments of Section 4.1.

In the case $d = n - 1$ this system gives no additional information to the property that $M_k f(y_0, \ldots, y_d)$ depends only on A. The system (5.8) is empty, if we take for instance, the point y_j on the $j + 1$th axis for $j = 0, \ldots, n-1$. In this case the moment conditions of Section 2.4 describe the range of M_{n-1}. For the opposite, in the case $d < n - 1$, (5.8) gives enough information. Consider the operator M_d defined on the Schwartz space $S(\mathbf{E})$. The range of this operator fulfils also the moment conditions (8.19) and the system of John equations. The latter gives a sufficient condition for a function to belong to the range of M_d:

Theorem 5.8 *If $d < n - 1$, then a function g defined in A_d belongs to the range of M_d on the Schwartz space $S(\mathbf{E})$, if and only if the function $g(y_0, \ldots, y_d) = g(A)$ is smooth, fast decreasing at infinity and satisfies* (5.8).

Theorem 5.9 *Any function g with compact support in A_d that fulfils* (5.8) *satisfies the moment conditions.*

This implies by Helgason's theorem that $g = M_d f$ for a function f with compact support (which is, of course, uniquely defined).

5.5 Duality for the Funk transform

We state a duality between the Funk transforms M_k and M_{n-1-k} on a sphere. This duality gives a method to translate an arbitrary reconstruction method for M_k to a reconstruction method for M_{n-1-k}. A similar relation holds in a Euclidean space of dimension n.

Let $\mathbf{V}$ be a vector space of dimension $n+1$ and $\mathbf{V}^*$ be the dual space. Choose a coordinate system $x_0, \dots, x_n$ in $\mathbf{V}$ and consider the even volume form $\mathrm{d}x = \mathrm{d}x_0 \wedge \cdots \wedge \mathrm{d}x_n$ in $\mathbf{V}$. Denote $\omega = \omega(\mathbf{V}) = \varepsilon \vdash \mathrm{d}x$ where ε is the Euler field in $\mathbf{V}$. Take an arbitrary homogeneous function f in $\mathbf{V}\setminus\{0\}$ of degree $-n-1$. The integral

$$\int_K f\omega$$

is well defined for an arbitrary n-chain $\mathbf{C}$ in $\mathbf{V}\setminus\{0\}$ and vanishes if $\mathbf{C} = \partial\mathbf{B}$ for an $n+1$-chain $\mathbf{B}$, since the form $f\omega$ is closed. Indeed, we have $\varepsilon(\mathrm{d}f) = -(n+1)f$ since f is a homogeneous function, $\mathrm{d}\omega = (n+1)\,\mathrm{d}x$, and $\varepsilon \vdash \mathrm{d}(f\omega) = \varepsilon(\mathrm{d}f)\,\omega + f\,\varepsilon \vdash \mathrm{d}\omega = -(n+1)f\,\omega + (n+1)f\omega = 0$. This implies $\mathrm{d}(f\omega)$ since this a form of highest degree.

Suppose for simplicity that $\mathbf{V}$ is in Euclidean space; note that the form $\omega(\mathbf{V})$ as above does not depend on the Euclidean structure, since the volume form $\mathrm{d}x$ does not. Take a subspace L in $\mathbf{V}$ of dimension k; let $\omega(L)$ be the corresponding $k-1$-form in L. Denote by $L^\perp$ the polar subspace of the dual space $\mathbf{V}^*$.

Theorem 5.10 *Let f be an arbitrary continuous homogeneous function in $\mathbf{V}\setminus\{0\}$ of degree $-k$, L be an arbitrary subspace of $\mathbf{V}$ of dimension $k \le n$ and $L^\perp$ be its polar in $\mathbf{V}^*$. We have*

$$\int_{\mathbf{S}(L^\perp)} \widehat{f}\,\omega\left(L^\perp\right) = \int_{\mathbf{S}(L)} f\,\omega(L), \tag{5.9}$$

where $\hat{f} = F(f\mathrm{d}x)$ and $\mathbf{S}(L)$ means the unit sphere in L.

Remark. According to the above remark, we can replace the sphere $\mathbf{S}(L)$ by any other homotopically equivalent cycle. The same is true for the cycle $\mathbf{S}\left(L^\perp\right)$.

◄ Let $k = n$; we apply Proposition 1.4, choosing $\Gamma = \mathbf{S}$ in (1.8). Take the sum for $\xi = \theta, -\theta$, where θ is a unit normal to L:

$$\widehat{f}(\theta) + \widehat{f}(-\theta) = \int_{\mathbf{S}} [\tau_{-1}(\langle\theta, x\rangle) + \tau_{-1}(-\langle\theta, x\rangle)] f(x)\omega(\mathbf{V}). \tag{5.10}$$

Meantime

$$\tau_{-1}(t) + \tau_{-1}(-t) = \mathrm{j}^{-1}[(t - 0\imath)^{-1} + (-t - 0\imath)^{-1}] = \delta_0(t)$$

and the right-hand side of (5.10) is equal to $\int_{\mathbf{S}} \delta_0(\langle \theta, x\rangle) f(x)\omega\,(\mathbf{V})$. This implies the equation

$$\widehat{f}(\omega) + \widehat{f}(-\omega) = \int_{\mathbf{S}(L)} f(x)\omega\,(L)\,. \tag{5.11}$$

The left side is equal to the integral of the function $\widehat{f}\omega\,(L^{\perp})$ over the intersection $\mathbf{S} \cap L^{\perp}$, hence (5.10) follows.

Let now $k < n$ and $\theta_0, \dots, \theta_p$, $p = n - k$ be an orthoframe in $L^{\perp}$. We apply the equation (5.11) for $\theta = \theta_0$ and for the product of a homogeneous generalized function $g(x) = f(x)\delta(\langle\theta_1, x\rangle) \cdot \dots \cdot \delta(\langle\theta_p, x\rangle)$ of degree $-n$:

$$\widehat{g}(\theta_0) + \widehat{g}(-\theta_0) = \int_{\mathbf{S}\cap L_0} g(x)\omega\,(L_0) = \int_{\mathbf{S}\cap L} f(x)\omega\,(L)\,, \tag{5.12}$$

where $L_0 = (\theta_0)^{\perp}$. Note that

$$\widehat{g}(\xi) = \int_{\mathbb{R}^p} \widehat{f}(\xi + t_1\theta_1 + \dots + t_p\theta_p)\,\mathrm{d}t_1 \wedge \dots \wedge \mathrm{d}t_p = \int_{\xi + G_0} f\,\mathrm{d}t,$$

where $G_0 = L^{\perp} \cap (\theta_0)^{\perp}$ is a cycle oriented by the form $\mathrm{d}t = \mathrm{d}t_1 \wedge \dots \wedge \mathrm{d}t_n$. The integral converges, since $\widehat{f}$ is homogeneous of degree $-p-1$. Therefore

$$\widehat{g}(\theta_0) + \widehat{g}(-\theta_0) = \int_G \widehat{f}\,\mathrm{d}t, \tag{5.13}$$

where $G = G_+ \cup G_-$, $G_\pm \doteq \pm\theta_0 + G_0$ is the cycle in $L^{\perp}$. We orientate the cycle $G_\pm$ by the form $\pm\mathrm{d}t$ and write the right-hand side of (5.13) in the form $\int_G \widehat{f}\omega_t$, where $\omega_t = e \dashv (\mathrm{d}t_0 \wedge \mathrm{d}t)$ is an n-form in G. This form coincides with $\omega\,(L^{\perp})$ on the cycle G and this cycle is homotopic to the sphere $\mathbf{S}^* \cap L^{\perp} = \mathbf{S}\,(L^{\perp})$. The equations (5.12) and (5.13) imply (5.9). ▶

5.6 Duality in Euclidean space

Definition. Let $\mathbf{E}_0$ be a Euclidean space of dimension n with the inner product $\langle , \rangle$. Take an affine subspace $A \subset \mathbf{E}_0 \backslash \{0\}$ of dimension k and consider the system of equations for $x \in \mathbf{E}_0$:

$$\langle x, y\rangle + 1 = 0,\ y \in A.$$

The set $\tilde{A}$ of solutions is an affine subspace $\mathbf{E}_0$ of dimension $n - k - 1$. We call this space *affine dual* to A. The double affine dual space to A coincides with A.

Let $M_{\mathbf{E}}$ be the flat integral transform in $\mathbf{E}_0$. It turns out that the values of $M_{\mathbf{E}}$ on A and $\tilde{A}$ are related as follows. Consider the standard embedding $e : \mathbf{E}_0 \to \mathbf{E}$ by $x \mapsto (1, x)$ where $\mathbf{E} = \mathbb{R} \dotplus \mathbf{E}_0$ is a Euclidean space. Fix an integer $k, 0 < k < n$; let f be a function in $\mathbf{E}_0$ such that

$$\left(1 + |x|^{2k+2}\right) f\,(x) \in L_1\,(\mathbf{E}_0)\,. \tag{5.14}$$

Define the function in $\mathbf{E}$,

$$g(x_0, x) \doteq x_0^{-k-1}\left(1+\left|\frac{x}{x_0}\right|^2\right)^{(k+1)/2} f\left(\frac{x}{x_0}\right).$$

It is homogeneous of degree $-k-1$. By (5.14) the density $g\mathrm{d}x_0 \wedge \mathrm{d}x$ is locally integrable and the Fourier transform $\widehat{g}$ is well defined in $\mathbf{E}^* = \mathbf{E}$. It is a homogeneous generalized function of degree $k-n$. We call

$$\tilde{f}(x) \doteq \left(1+|x|^2\right)^{(k-n)/2} \widehat{g}(1, x)$$

a k-dual function to f. It is easy to see that the function $n-k-1$-dual to $\tilde{f}$ is equal to f provided that $\left(1+|x|^{2n-2k}\right)\tilde{f}(x) \in L_1(\mathbf{E}_0)$. Denote $\mathrm{d}(A) \doteq \mathrm{dist}(A, 0)$.

Theorem 5.11 *Let f be a function in $\mathbf{E}_0$ satisfying (5.14) for some integer k, $0 < k < n$. Then for an arbitrary affine subspace $A \subset \mathbf{E}_0$ of dimension k we have*

$$\mathrm{d}^{1/2}\left(\tilde{A}\right) M\tilde{f}\left(\tilde{A}\right) = \mathrm{d}^{1/2}(A)\, M f(A), \tag{5.15}$$

where $\tilde{f}$ is the k-dual function.

◀ Let $\mathbf{S}_+ \doteq \left\{x_0^2 + |x|^2 = 1, x_0 > 0\right\}$ and L be the linear envelope of $e(A) \cup \{0\}$ in $\mathbf{E}$. The form $g\omega(\mathbf{E})$ is integrable in $\mathbf{S}_+$ and by (3.8) we have

$$\begin{aligned}\int_{\mathbf{S}_+ \cap L} g\omega(\mathbf{E}) &= \left(1+d^2(A)\right)^{1/2} \int_A \left(\left(1+|x|^2\right)^{-(d+1)/2} g(1,x)\right) \mathrm{d}V(A) \qquad (5.16)\\ &= \left(1+d^2(A)\right)^{1/2} M_{\mathbf{E}} f(A).\end{aligned}$$

On the other hand, the function g is defined in $\mathbf{E}$ as a homogeneous function of degree $-k-1$. Therefore by Theorem 5.10 we have

$$\int_{\mathbf{S}_+ \cap L} g\omega(L) = \frac{1}{2}\int_{\mathbf{S}(L)} g\omega(L) = \frac{1}{2}\int_{\mathbf{S}(L^\perp)} \widehat{g}\omega\left(L^\perp\right) = \int_{\mathbf{S}_+ \cap L^\perp} \widehat{g}\omega\left(L^\perp\right).$$

The polar space $L^\perp$ is equal to the envelope of $\tilde{A} \cup \{0\}$. We apply the equation (5.16) to the right side:

$$\int_{\mathbf{S}_+ \cap L^\perp} \widehat{g}\omega\left(L^\perp\right) = \left(1+d^2\left(\tilde{A}\right)\right)^{1/2} M_{\mathbf{E}}\tilde{f}\left(\tilde{A}\right). \tag{5.17}$$

The formula (5.15) follows from (5.16) and (5.17), if we take into account that $d(A)\, d\left(\tilde{A}\right) = 1$. ▶

Example 1. Let Γ be a smooth curve in $\mathbb{P}^3$ and $\Lambda(\Gamma)$ the variety of lines that meet Γ. The dual variety $\tilde{\Lambda}(\Gamma)$ is equal to the variety $\Lambda(S)$ of lines $\tilde{\lambda}$ in $\tilde{\mathbb{P}}^3$ that are tangent to a surface S. The latter is naturally isometric to the tangent surface $T(\Gamma)$ of the curve Γ which is the union of all tangent lines to Γ, This yields

$$\tilde{\Lambda}(\Gamma) \cong \Lambda(T(\Gamma)).$$

The variety $T(\Gamma)$ is a *ruled* surface, i.e., it has an isometric embedding in a plane. Lines of $T(\Gamma)$ correspond by the projective duality to points in Γ.

Example 2. Let Γ be a smooth curve in $\mathbf{E}^n \backslash \{0\}$ and $\Lambda(\Gamma)$ the variety of lines that meet Γ. The dual variety $\tilde{\Lambda}(\Gamma)$ consists of affine $n-2$-planes A that are contained in hyperplanes $\tilde{\gamma}$, $\gamma \in \Gamma$ where $\tilde{\gamma}$ is dual to the point γ. The family $\{\tilde{\gamma}, \gamma \in \Gamma\}$ has the envelope S which is a smooth hypersurface in $\mathbf{E}$, if Γ generic. It is true, for instance, if the vectors $x'(s), x''(s), \dots, x^{(n)}(s)$ are independent at each point $x = x(s)$ of Γ. The variety $\tilde{\Lambda}(\Gamma)$ is the family of $n-2$-planes that are tangent to S. The hypersurface S has, of course, a very special form. For the variety $\Lambda(\Gamma)$ we can apply the reconstruction formula of Theorem 4.8. By Euclidean duality we get a reconstruction method for the variety $\tilde{\Lambda}(\Gamma)$.

Example 3. Let S be a surface in $\mathbf{E}^4$ with non-vanishing Gaussian curvature and $\Lambda(S)$ the variety of lines tangent to S. Then we have

$$\tilde{\Lambda}(S) = \Lambda(\tilde{S}),$$

where $\tilde{S}$ is the dual surface, i.e., the envelope of hyperplanes $\lambda(x)^{\perp}$, $x \in S$, where $\lambda(x)$ denotes the line through x and the origin in $\mathbf{E}^4$.

Chapter 6

Incomplete Data Problems

6.1 Completeness condition

Let (X, g) be a Riemannian manifold and Y be a variety of closed submanifolds $Y \subset X$. Consider the integral transform for the variety Y:

$$Mf(Y) = \int_Y f \, \mathrm{d}V(Y), \; Y \in \mathsf{Y}, \tag{6.1}$$

where $\mathrm{d}V(Y)$ is the volume element on Y induced by the metric g. The reconstruction problem is to find the function f from data of $Mf|\mathsf{Y}$. More complicated versions of (6.1) arise in applications. A weight function $w = w(x, Y)$ (known or unknown) can appear in the integral, the "image" f can be a section of a tensor bundle, like a differential symmetric or skew symmetric form.

We focus on the simplest case where f is a scalar function. A closed analytic reconstruction formula is known in only a few cases. If there is no such formula one can try to apply numerical methods. An actual numerical algorithm contains usually a regularization procedure and gives a convergent result whatever the input data are. To ensure reliability of the result, the family Y (i.e., the acquisition geometry) should be big enough to guarantee existence of a continuous reconstruction operator $R : Mf|\mathsf{Y} \mapsto f$. The condition of continuity can be specified for a family Y that has the structure of a smooth manifold. The mapping R is then supposed to be continuous as an operator from the space of smooth functions f with compact support to a space of smooth functions in Y.

The completeness condition gives an answer to this question. Let X be the space where an unknown original function f is defined. We wish to reconstruct f from the mean transform Mf defined for a family Y of submanifolds of X.

Definition. Let Y be a family of submanifolds of X. It satisfies the *completeness condition* in an open subset $U \subset X$, if for an arbitrary $x \in G$ and arbitrary covector t at x there exists $Y \in \mathsf{Y}$ such that $x \in Y$ and t is conormal to Y. In other words, the union of conormal bundles $N^*(Y)$ for $Y \in \mathsf{Y}$ contains the cotangent bundle $T^*(U)$. This condition is necessary for reconstruction: if the family Y is endowed with the structure of a smooth manifold and there exists a reconstruction operator

R defined on the space of test functions $D(U)$ that is continuous in C^∞-topologies, then the family Y fulfills the completeness condition in U.

On the other hand, if a variety Y is complete in K, then there exists, at least, a quasi-inverse operator for R. Here we use the terminology of the theory of pseudodifferential operators (PDO), see, e.g., [92]. Apply the backprojection operator

$$M^\sharp g\,(x) = \int_{\mathsf{Y}(x)} g\,(Y)\,\mathrm{d}\sigma \tag{6.2}$$

to a function g defined on the variety Y_x of manifolds $Y \in \mathsf{Y}$ that contain a point x. Here $\mathrm{d}\sigma$ is a measure on this variety. The completeness condition for the generic family Y implies that the operator $A = M^\sharp M$ is a pseudodifferential operator (PDO) of elliptic type. Applying an arbitrary pseudodifferential operator B that is quasi-inverse of A, we get a quasi-reconstruction $f + Sf = BM^\sharp g$, where S is a PDO of order ≤ -1.

Example. By Proposition 2.3 for the Radon transform $M = R$ we have

$$R^\sharp Rf\,(x) = \frac{\pi^{(n-1)/2}}{\Gamma\,((n-1)\,/2)} \int \frac{f\,(y)}{|x-y|}\mathrm{d}y.$$

This is a PDO with the symbol $|\xi|^{1-n}$. By (2.7) the PDO $c\Delta^{(n-1)/2}$ is the exact inverse for a constant c.

In most practical situations the set of available projections is incomplete. No inversion problem with incomplete data can be solved by means of an explicit formula or by a stable numerical algorithm. Any numerical reconstruction is based in this case on some regularization method. The natural question arises: what is the similarity of the regularized solution to the real object whose projection data were used. A partial answer can be found in the phase space terms which means localization of a function in the cotangent bundle $T^*\,(X)$.

Suppose that a compact set $K \subset X$ is known such that $\operatorname{supp} f \subset K$. If no more a priori information is accessible, the energy of unknown original f is assumed to be spread uniformly over the cotangent bundle $T^*(K)$. Take a manifold $Y \in \mathsf{Y}$ and consider the conormal bundle $N^*(Y \cap K) \subset T^*(K)$ of this curve. Denote by $A(\mathsf{Y})$ the union of sets $N^*(Y \cap K)$. This is a conic subset of $T^*(K)$. We call this subset the *audible* zone. It can be shown that the part of the energy of the original f inside the audible zone can be reasonably estimated by a suitable norm of Mf. The complementary part of the energy which is contained in the *silent* zone $T^*(K)\backslash N^*(\mathsf{Y})$ can be estimated with a weight. This weight is a function in the cotangent bundle that exponentially decreases, when the point moves away from the audible zone. We shall see below (6.5) an example of an estimate of this kind.

6.2 Radon transform of Gabor functions

Now we clarify the meaning of the audible zone in a quite simple model. Let $\mathbf{E}$ be a Euclidean space of dimension n. Choose a unit of length $\sigma > 0$. A function of the form

$$e_\lambda(x) = e_{\lambda,\sigma}(x) \doteq (2\sigma)^{n/4} \exp\left(-\pi\sigma^2 (x-q)^2 + \jmath\theta x\right) \tag{6.3}$$

is called a *Gabor* function. Here $\lambda = (p,\theta)$ is an arbitrary point in the phase space $\mathbf{E} \times \mathbf{E}^*$. The phase space has the natural Euclidean structure: $\|\lambda\|_\sigma^2 = \sigma^2 |q|^2 + \sigma^{-2} |\theta|^2$. We shall use the notation $\langle \cdot, \cdot \rangle$ for the scalar product in $L_2(X)$ and $\|\cdot\|$ for the norm. Note that for arbitrary points $\lambda = (q,\theta), \mu = (r,\eta) \in \mathbf{E} \times \mathbf{E}^*$ we have

$$\langle e_\lambda, e_\mu \rangle = \exp(\pi\imath (q+r)(\theta-\eta)) \exp\left(-\pi |\lambda-\mu|^2/2\right), \ \langle e_\lambda, e_\lambda \rangle = 1 \tag{6.4}$$

and

$$\hat{e}_\lambda = e_{\hat{\lambda}}, \ \lambda = (q,\theta), \ \hat{\lambda} = (\theta, -q).$$

We evaluate the integral of e_λ along the hyperplane $H(p,\omega)$. Choose Euclidean coordinates in such a way that $\omega = (1,0,\dots,0)$. Then we have

$$\int_{H(p,\omega)} e_\lambda(x)\,\mathrm{d}S = (2\sigma)^{1/4} \exp\left(-\pi\sigma^2 \left[(p-\omega q)^2 + |\theta_\omega|^2\right]\right) \exp\left(\jmath\left(p\langle\omega,\theta\rangle + \theta_\omega q\right)\right),$$

where $\theta_\omega \doteq \theta - \langle \omega, \theta \rangle \omega$ is the projection of θ to the plane orthogonal to ω. This equation yields

$$(2\sigma)^{-1/4} |R[e_\lambda](p,\omega)| = \exp\left(-\pi\left[\sigma^2 (p-\omega q)^2 + \sigma^{-2} |\theta_\omega|^2\right]\right).$$

The right-hand side is equal to 1 if $p = \omega q$, $\theta_\omega = 0$, which means that the center of e_λ belongs to $H(p,\omega)$ and θ is collinear to ω. Otherwise this quantity decreases exponentially. The result can be interpreted as follows. Define the distance function d_σ in $\mathbf{E} \times \mathbf{E}^*$ generated by the norm $\|\cdot\|_\sigma$ as above.

Proposition 6.1 *For an arbitrary hyperplane H and a point $\lambda = (q,\theta)$ the equation*

$$|R[e_\lambda](H)| = (2\sigma)^{1/4} \exp\left(-\pi d_\sigma^2 (\lambda, N^*(H))\right)$$

holds, where d is the distance in $\mathbf{E} \times \mathbf{E}^*$ and $N^*(H)$ is the conormal bundle of H, i.e., the set of all conormal vectors to H.

6.3 Reconstruction from limited angle data

Proposition 6.1 implies that the function $f = e_\lambda$ cannot be stably reconstructed if θ is not in the audible angle diapason, precisely if $d(\theta, A)$ is big enough where $A \subset T^*(X)$ is the audible zone. We discuss now a special case of this result in more detail. Let f be a function in a Euclidean space $\mathbf{E}$; suppose that the Radon transform $Rf(p,\omega)$ is known only for ω in an open set $\Omega \subset \mathbf{S}^{n-1}$ and all $p \in \mathbb{R}$. This means that the audible zone is just $\mathbf{E} \times L(\Omega)$ where $L(\Omega) \doteq$

$\{\xi = t\omega; \omega \in \Omega, t \in \mathbb{R}\}$ is the union of lines in $\mathbf{E}^*$ spanned by Ω. The function $\hat{f}$ is known in $L(\Omega)$ and can be interpolated to X^* if f has compact support.

For simplicity we consider the case $\dim X = 2$. Let Ω be the set of unit vectors ω whose angles ϕ with x_1-axes fulfils the inequality $\tan\phi \leq t < \infty$. We can interpolate $\hat{f}$ in the silent zone $\mathbf{E} \times \mathbf{E}^* \setminus L(\Omega)$ by means of Corollary 1.23. In the silent zone only a weak estimate holds. Consider the quadratic function $q(\xi) \doteq \xi_1^2 - d^2 |\xi'|^2$. It is positive in the audible zone A and negative in the silent zone.

Theorem 6.2 *For any function $f \in L_2(X)$ whose support is contained in the strip $\{|x_1| \leq a\}$ the inequality holds,*

$$\int_{q<0} |\exp\left(-2\pi a\sqrt{-q(\xi)}\right) \hat{f}(\xi)|^2 \mathrm{d}\xi \leq \int_{q\geq 0} |\hat{f}(\xi)|^2 \mathrm{d}\xi. \tag{6.5}$$

◀ We have $F_{p\to\rho}Rf = \hat{f}(\rho\omega)$, hence we know the Fourier transform $\hat{f}(\xi)$ in the domain $A = \{|\eta| \leq d|\xi_1|\}, \eta = (\xi_2, \ldots, \xi_n)$, $d = \tan\alpha$ (a spherical cone around the x_1-axes). Fix η-coordinates and consider the function $\phi(\zeta) \doteq \hat{f}(\zeta, \eta)$. It is a-bandlimited and is known for $|\zeta| \geq d|\eta|$. Applying Corollary 1.23 yields

$$\int \left|\exp\left(-2\pi a\sqrt{d^2 |\eta|^2 - \xi_1^2}\right) \hat{f}(\xi)\right|^2 \mathrm{d}\xi_1 \leq \int_{q\geq 0} |\hat{f}(\xi)|^2 \mathrm{d}\xi_1.$$

Integrating this inequality against $\mathrm{d}\eta$, gives (6.5). ▶

This estimate cannot be much improved. This follows from Proposition 6.1.

6.4 Exterior problem

The problem is to reconstruct a function f in Euclidean plane $\mathbf{E}^2$, from knowledge of line integrals for lines $L \subset \mathbf{E}\backslash B$ where B is the unit ball. There is no simple reconstruction formula. The solution given by A. Cormack is based on the decomposition of f and $M_1 f$ in harmonics. Let K be a convex set in X. How can one reconstruct a function $f \in S(\mathbf{E})$ in the complement $\mathbf{E}\backslash K$ from knowledge of the integrals $Rf(H)$ for hyperplanes $H \subset \mathbf{E}\backslash K$? This question is called the exterior problem for the Radon transform. The uniqueness in the exterior problem holds if K is compact:

Theorem 6.3 *If a function $f \in S(\mathbf{E})$ satisfies the equation $Rf(H) = 0$ for any hyperplane $H \subset \mathbf{E}\backslash K$, where K is a compact convex set in $\mathbf{V}$, then* $\operatorname{supp} f \subset K$.

This fact follows from Theorem 6.5 for the unit ball $B = \{x, |x| \leq 1\}$. It is also true for an arbitrary convex compact set K, since for any point $x \in V \setminus K$ there is a ball B'such that $K \subset B' \subset V\backslash\{x\}$.

Spherical harmonics. A function h in V is called harmonic if it satisfies the Poisson equation $\Delta h = 0$. A function $\phi(\omega)$ on the unit sphere $\mathbf{S} = \mathbf{S}^{n-1}$ is called a spherical harmonic of degree $k = 0, 1, 2, \ldots$, if there exists a homogeneous harmonic polynomial h of degree k such that $\phi = h|\mathbf{S}$. For any k the space of spherical harmonics of degree k is of dimension $d(n,k) = (n+k-3)!(n+2k-2)\ /(n-2)!k!$. Any spherical harmonic of degree k is orthogonal to an arbitrary spherical harmonic of degree $l \neq k$ with respect to the spherical scalar product

$$\langle a, b\rangle_S \doteq \int_{\mathbf{S}} a\bar{b} \mathrm{d}S.$$

The space of square integrable functions on the sphere is denoted by $L_2(\mathbf{S})$. The system of spherical harmonics is complete in $L_2(\mathbf{S})$ and one can choose a countable orthobasis $\{\phi_\alpha, \alpha = 0, 1, \ldots\}$ consisting of spherical harmonics. The basic fact of the harmonic analysis on the sphere is

Theorem 6.4 *Any function $\psi \in L_2(\mathbf{S})$ can be developed in a series of harmonics*

$$\psi(\omega) = \sum \langle \psi, \phi_\alpha \rangle \phi_\alpha.$$

This series is unique and converges to ψ in $L_2(\mathbf{S})$.

This series is called the harmonic decomposition of the function.

Orthogonal polynomials. Take an arbitrary real number $\lambda > -1$ and the measure $\mathrm{d}\sigma \doteq \left(1 - t^2\right)^{\lambda - 1/2} \mathrm{d}t$ on the interval $[-1, 1] \subset \mathbb{R}$. Consider the Hilbert space $L_2(\mathrm{d}\sigma)$ of functions in the interval that are square-integrable with respect to this measure. The system of orthogonal polynomials in $L_2\,(\mathrm{d}\sigma)$,

$$G_k^\lambda = c(\lambda, k)(1-t^2)^{-\lambda+1/2} \frac{\mathrm{d}^k}{\mathrm{d}t^k} (1-t^2)^{\kappa+\lambda-1/2}, \ k = 0, 1, \ldots,$$

normalized by the condition $G_k^\lambda(1) = 1$, are called Gegenbauer polynomials (a special case of Jacobi polynomials). For $\lambda = 0$ they coincide with the Chebyshev polynomials of first kind: $G_k^0(t) = T_k(t) = \cos\left(k \arccos t\right)$.

Reconstruction in the complement to the ball. Take a function $f \in S(\mathbf{V})$ and consider for each $r > 0$ the function $\psi_r(\omega) \doteq f(r\omega)$ on the unit sphere. Apply the harmonic decomposition

$$f(r\omega) = \sum_\alpha \langle \psi_r, \phi_\alpha \rangle \phi_\alpha(\omega).$$

The function $f_\alpha(r, \omega) = \langle \psi_r, \phi_\alpha \rangle \phi_\alpha(\omega)$ is for any α and r a harmonic polynomial of degree k in ω. For the Radon transform $g = Rf$ we get a similar equation

$$g(p, \omega) = \sum g_\alpha(p, \omega).$$

Theorem 6.5 *For any r and α we have*

$$f_\alpha(r,\omega) = \frac{(-1)^{n-1}}{2^{n-2}\pi^\lambda}\sum_\alpha \int_1^\infty \frac{(t^2-1)^{\lambda-1}}{\Gamma(\lambda)} G_{k(\alpha)}^{\lambda-1/2}(t) g_\alpha^{(n-1)}(tr,\omega)\mathrm{d}t,$$

where $\lambda \doteq (n-1)/2$ and $k(\alpha)$ is the degree of the spherical harmonic h_α.

In particular, for $n = 2$ the kernel in the integral is the Chebyshev polynomial of order k. The Chebyshev polynomials exponentially grow to infinity in any point $t > 1$ as $\lambda \to \infty$. So do the Jacobi polynomials for any λ. Therefore the reconstruction given by this theorem is exponentially unstable. Nevertheless, it implies a uniqueness result: if $g(p,\omega) = 0$ for $p > 1$ and any ω, then $f(r\omega) = 0$ for $r > 1$, q.e.d.

STABILITY. The audible zone $A(Y)$ in the exterior problem is the union of conormal bundles $N^*(H)$ of all hyperplanes $H \subset \mathbf{E}\backslash K$. For a point $q \in \mathbf{E}\backslash K$ the fibre $A_q(Y)$ is close to X^* if q is far from K. On the other hand, if q is close to K the set $A_x(Y)$ is a very narrow cone. This means that any reconstruction is very unstable in q. To make this assertion more quantitative we choose a sequence of functions f which behave as "ghosts", i.e., the exterior Radon data of f are very small. We take Gabor functions (6.3) for the test; note that $\|e_{\lambda,\sigma}\|_{L_2} = 1$ for any λ. We fix a number $r > 0$ and choose the scaling parameter σ such that $\sigma d = r$ where $d = d_X(q,K)$ and d_X is the Euclidean distance in $\mathbf{E}$. The "detail" described by a Gabor function $e_{\lambda,\sigma}, \lambda = (q,\theta)$ is almost supported in $\mathbf{E}\backslash K$ since $|e_{\lambda,\sigma}| \leq (2\sigma)^{n/4}\exp(-\pi r^2)$ is small in K. Take any θ orthogonal to $\nu(y)$. By Proposition 6.1 we have

$$|R[e_{\lambda,\sigma}](H)| = (2\sigma)^{n/4}\exp\left(-\pi\min\left(\sigma^2 d^2(q,H) + \sigma^{-2}|\theta_\omega|^2\right)\right),$$

where the minimum is taken over all hyperplanes $H \subset \mathbf{E}\backslash K$; ω is the normal vector. Let $\Omega(q)$ be the set of normal vectors ω for hyperplanes $H \subset \mathbf{E}\backslash K$ such that $d(q,H) \leq d/\sqrt{2}$. If the point q is close to a smooth point $y \in \partial K$, then $\Omega(q)$ is contained in an ε-neighborhood of the line normal to ∂K at y and $\varepsilon \leq 3\sqrt{\kappa d}$ where κ is the minimal normal curvature of ∂K at y. Then $|\theta_\omega|^2 \geq \left(1-\sin^2\varepsilon\right)|\theta|^2 \geq (1-9\kappa d)|\theta|^2$. Thus for $H \in \Omega(q)$ we have the estimate $\sigma^{-2}|\theta_\omega|^2 \geq (1-9\kappa d)\sigma^{-2}|\theta|^2 \geq \sigma^{-2}|\theta|^2/2 =$ for $18\kappa d \leq 1$. In the opposite case we have $\sigma^2 d^2(q,H) \geq (\sigma d(q,H))^2 \geq (\sigma d)^2/2 = r^2/2$. This yields

$$\sigma^2 d^2(q,H) + \sigma^{-2}|\theta_\omega|^2 \geq \min\left(r^2, |d(q,K)\theta/r|^2\right)/2$$

for $d \leq R/18$ where R is the minimal curvature radius of ∂K and any hyperplane H. To optimize the minimum, we take $|\theta| = r^2/d(q,K)$ and get

$$\max_{H\subset X\backslash K}|R[e_{\lambda,\sigma}](H)| \leq (2\sigma)^{n/4}\exp\left(-\frac{\pi}{2}r^2\right),\ d = d(q,K). \tag{6.6}$$

Corollary 6.6 *Suppose that K is a convex compact in X with smooth boundary. For an arbitrary point $q \in \mathbf{E} \backslash K$ sufficiently close to K and an arbitrary $r > 0$ the estimate (6.6) holds for $\sigma = r/d$ and $\lambda = (q, \theta)$ where θ is an arbitrary vector in the tangent plane $T_y(\partial K)$ at the nearest point y to q, $|\theta| = r^2/d$.*

6.5 The parametrix method

If integral data are complete, one can construct a quasi-inverse operator explicitly for a rather wide class of geometries. We follow here Beylkin's method. Let $X \subset V = \mathbb{R}^n$ be an open set and $\phi : X \times \mathbf{S}^{n-1} \to \mathbb{R}$ be a C^{n+1}-smooth function that fulfils the condition $(*)$: *the equation*

$$\theta = t\phi'_x(x, \omega) \tag{6.7}$$

has the unique solution $t = t(x, \theta) > 0, \omega = \omega(x, \theta)$ for $(x, \theta) \in X \times \mathbb{R}^n \backslash \{0\}$; this solution is smooth in x, θ.

The set $H(s, \omega) \doteq \{x \in X; \phi(x, \omega) = s\}$ is a smooth hypersurface. The condition $(*)$ implies the completeness condition for the variety of manifolds $H(s, \omega)$. These manifolds play now the role of hyperplanes in the affine geometry or big spheres in the spherical geometry.

Embed $\mathbf{S}^{n-1}$ as a unit sphere in a Euclidean space $\mathbb{R}^n$ with coordinates $\theta = (\theta_1, \dots, \theta_n)$ and extend the phase function on $X \times \mathbb{R}^n \backslash \{0\}$ to the homogeneous function $\phi(x, \theta) \doteq |\theta| \phi(x, \theta/|\theta|)$. Define the function

$$h(x, \theta) = \det \left\{ \frac{\partial^2 \phi(x, \theta)}{\partial x_j \partial \theta_k} \right\}_{j,k=1}^{n}.$$

This determinant does not vanish due to (ii). Fix a coordinate system $x_1, \dots, x_n$ and consider the volume form $\mathrm{d}x = \mathrm{d}x_1 \wedge \dots \wedge \mathrm{d}x_n$ in V. Take a smooth positive function a in $X \times \mathbf{S}^{n-1}$; define the generalized Radon-Funk transform by the integral

$$Mf(s, \omega) \doteq \int_{H(s,\omega)} a(x, \omega) f(x) \frac{\mathrm{d}x}{\mathrm{d}\phi},$$

where the orientation of $H(s, \omega)$ is defined by the form $\mathrm{d}\phi$. The integral Mf coincides with the Funk and the Radon transform when ϕ is a linear function of the point x running over a sphere X or a Euclidean space X, respectively. To reconstruct function f from the data Mf we follow the method of Theorem 2.9. For this we apply the standard pseudodifferential operator with the symbol t_+^{n-1} with respect to the variable s:

$$Tg(s) = F^*_{t \to s}\left(t_+^{n-1} F_{s \to t}(g)\right).$$

Define the backprojection operator of the form (6.2) as

$$M^\sharp g(x) = \int_{\mathbf{S}^{n-1}} g(\phi(x, \omega), \omega)\, \mathrm{d}\sigma, \ g = g(s, \omega),$$

where we set $\mathrm{d}\sigma = b(x,\omega)\,\mathrm{d}\omega$, $\mathrm{d}\omega$ is the Euclidean area in the unit sphere $\mathbf{S}^{n-1}$ and b is a smooth function to be specified later.

Theorem 6.7 *The operator*

$$Qf = M^{\sharp}TMf$$

defined on the space $D(X)$ *is a pseudodifferential operator of order* 0 *with the principal symbol*

$$\sigma_Q(x,\omega) = \frac{a(x,\omega)\,b(x,\omega)}{|h(x,\omega)|}, \omega \in \mathbf{S}^{n-1}.$$

Remark 1. If we choose a and b in such a way that $a(x,\omega)\,b(x,\omega) = |h(x,\omega)|$, then we have $Q = I + S$ where I is the identity operator and S is a PDO of order -1. It means that the function $M^{\sharp}Tg$, where $g = Mf$ is a reconstruction of f up to the term Sf which is "smoother" than f. It follows that the functions $M^{\sharp}Tg$ and f have, at least, the same geometry of singularities.

Remark 2. An invariant definition of the function h is $h\mathrm{d}x\mathrm{d}\theta = \wedge^n \mathrm{d}_x\mathrm{d}_\theta\phi(x,\theta)$.

◀ We have $\mathrm{d}\omega = |\theta|^{1-n}\,\mathrm{d}\theta$, where $\mathrm{d}\theta = \mathrm{d}\theta_1 \wedge \cdots \wedge \mathrm{d}\theta_n$ and $\omega = \theta/|\theta|$. Write

$$\begin{aligned}
Qf(x) &= \int_{x\in H(s,\omega)} b(x,\omega)\,\mathrm{d}\omega \\
&\qquad \times \int\!\!\int_0^\infty \exp(\mathrm{j}(s-\sigma)t)\,|t|^{n-1}\,\mathrm{d}t\mathrm{d}\sigma \int_{H(\sigma,\omega)} a(y,\omega)\,f(y)\,\frac{\mathrm{d}y}{\mathrm{d}\phi} \\
&= \int_{\mathbf{S}} b(x,\omega)\,a(x,\omega)\,\mathrm{d}\omega\,\exp(\mathrm{j}t\,\phi(x,\omega))\,t_+^{n-1}\mathrm{d}t \int_X \exp(-\mathrm{j}t\,\phi(y,\omega))\,f(y)\,\mathrm{d}y \\
&= \int_{\mathbb{R}^n} b(x,\omega)\,a(x,\omega)\exp(\mathrm{j}\,\phi(x,\theta))\,\mathrm{d}\theta \int_X \exp(-\mathrm{j}\,|\theta|\,\phi(y,\omega))\,f(y)\,\mathrm{d}y,
\end{aligned}$$

where we set $t = |\theta|$. Substitute $f = F^*\hat{f}$, $\hat{f} = Ff$ and apply Fubini's theorem:

$$\begin{aligned}
Qf(\xi) &= \int_{\mathbb{R}^n}\int_X\int_{\mathbb{R}^n} b(x,\omega)\,a(x,\omega)\exp(\mathrm{j}\,[\xi y + \phi(x,\theta) - \phi(y,\theta)])\,\mathrm{d}\theta\mathrm{d}y\,\hat{f}(\xi)\,\mathrm{d}\xi \\
&= \int_{\mathbb{R}^n}\left(\int_X\int_{\mathbb{R}^n} b(x,\omega)\,a(x,\omega)\exp(\mathrm{j}\Phi(x,y,\theta,\xi))\,\mathrm{d}\theta\mathrm{d}x\right)\hat{f}(\xi)\,\mathrm{d}\xi.
\end{aligned}$$

Apply the stationary phase method to the inner integral. The critical points of the phase function $\Phi(x,y,\theta,\xi) = \xi y + \phi(x,\theta) - \phi(y,\theta)$ are given by the equations $\xi = \phi'_y(y,\theta)$, $\phi'_x(x,\theta) = \phi'_y(y,\theta)$. According to $(*)$ the last one is equivalent to $x = y$. We have

$$\Phi''_{y\theta} = \begin{pmatrix} \phi''_{yy} & \phi''_{y\theta} \\ \phi''_{\theta y} & \Phi''_{\theta\theta} \end{pmatrix}, \ \Phi''_{\theta\theta} = \phi''_{\theta\theta}(x,\theta) = \phi''_{\theta\theta}(y,\theta) = 0,$$

which yields $\det\Phi'' = (-1)^n \left|\det\phi''_{x\theta}\right|^2 = (-1)^n h^2$ and

$$\int_X \int_{\mathbb{R}^n} b(x,\omega)\, a(x,\omega) \exp\left(\mathrm{j}\Phi(x,y,\theta,\xi)\right) \mathrm{d}\theta \mathrm{d}x$$
$$= \left[\frac{a(x,\omega)\, b(x,\omega)}{|h(x,\omega)|} + s(x,\theta)\right] \exp\left(\mathrm{j}\xi x\right), \tag{6.8}$$

where s is a smooth function in $X \times \mathbb{R}^n \backslash \{0\}$ such that

$$D_x^\beta D_\theta^\alpha s(x,\theta) = O\left(|\theta|^{-1-|\alpha|}\right)$$

for big $|\theta|$ and any α, β such that $|\alpha| + |\beta| \leq n$. Note that the quadratic form related to the matrix Φ'' has signature (n,n), which makes no extra phase term appear in (6.8). Finally

$$Qf(x) = \int \exp\left(\mathrm{j}\langle\xi,x\rangle\right) \frac{a(x,\omega)\, b(x,\omega)}{|h(x,\omega)|} \hat{f}(\xi)\, \mathrm{d}\xi + \int \exp\left(\mathrm{j}\langle\xi,x\rangle\right) s(x,\theta)\, \hat{f}(\xi)\, \mathrm{d}\xi,$$

where ω and $\theta = t\omega$ are to be found from the equation (6.7). The first term is the PDO of order 0 with the symbol σ_Q and the second term is a PDO of order ≤ -1. ▶

Example 1. The function $\phi(x,\theta) = \langle x,\theta\rangle \doteq \sum x_j\theta_j$ fulfils condition $(*)$. It defines the hyperplane geometry and the above theorem follows from (2.9).

Example 2. We can take $\phi(x,\theta) = \rho(x)\langle x,\theta\rangle$ where $\rho(x) > 0$. We have then $\partial^2\phi/\partial x_j\partial\theta_k = \rho\delta_{jk} + x_j\rho'_k$ where δ_{jk} is Kronecker's. The assumption $(*)$ is obviously fulfilled, at least, in a neighborhood of the origin.

Chapter 7

Spherical Transform and Inversion

7.1 Problems

The spherical integral transform of a function f in a Euclidean space $X = \mathbf{E}^n$ is the family of the integrals

$$Mf(S) = \int_S f \, \mathrm{d}S, \tag{7.1}$$

where S is a sphere in $\mathbf{E}^n$ and $\mathrm{d}S$ denotes the Euclidean $n-1$-surface element. We shall also write $Mf(a,r) = Mf(S(a,r))$ where $S(a,r)$ denotes the sphere with the centre a and radius r. Replacing $\mathbf{E}^n$ by a Euclidean sphere $\mathbf{S}^n$, we define the spherical transform $M_{\mathbf{S}}f$ on the variety of spheres in $\mathbf{S}^n$. The reconstruction problems for $\mathbf{E}^n$ and $\mathbf{S}^n$ are equivalent if all hyperplanes in $\mathbf{E}^n$ are included as spheres of infinite radius.

The general reconstruction problem is to reconstruct a function f from knowledge of spherical integrals Mf on an n-dimensional subvariety (pencil) Σ of spheres. We focus on three special cases:

I. Σ is the pencil of spheres orthogonal to a hyperplane $\mathbf{H} \subset \mathbf{E}$;

II. the set of spheres with centres at a sphere and

III. Σ is the family of spheres tangent to a subvariety.

To allow the geometry to fulfill the completeness condition, we also include a set of hyperplanes to the pencil Σ in the cases **I** and **III**.

7.2 Arc integrals in the plane

Let $\mathbf{E}_+$ be an open half-plane in a Euclidean plane and Σ be the family of circle arcs and straight lines A in $\mathbf{E}_+$ that are orthogonal to $\mathbf{H} = \partial\mathbf{E}_+$, see Fig. 7.1.

The problem of inversion of the integral transform for the pencil Σ is a complete data problem in $\mathbf{E}_+$, since for any point $e \in \mathbf{E}_+$ and any tangent vector t in e there exists, at least, one curve $A \in \Sigma$ through e that is orthogonal to t. This condition is not satisfied in applications, since the integrals over very long arcs are not available. We consider the arc integral transform for limited data.

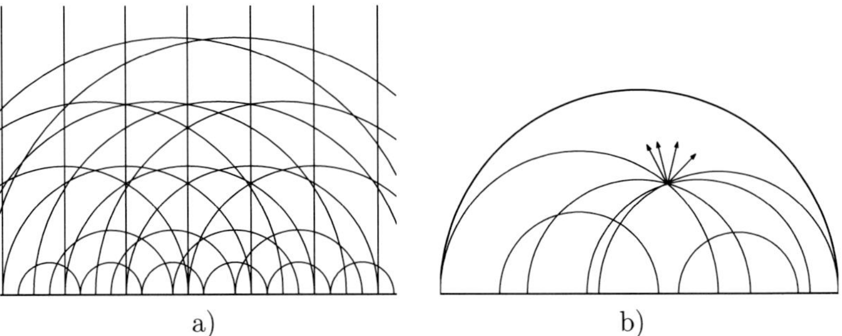

Figure 7.1: a) Arcs orthogonal to the x axis b) Arcs in the half-disc

Take the closed unit disc D in $\mathbf{E}^2$ and consider the half-disc $D_+ = \mathbf{E}_+ \cap D$. Let Σ_D be the subset of Σ consisting of arcs $A \subset D$. The problem is to reconstruct a function f in $\mathbf{E}$ with support $\operatorname{supp} f \subset D_+$ from the limited arc transform $Mf|\Sigma_D$. This is a problem with incomplete data since for any point $x \in D_+$ the normal vector t to arcs $A \in \Sigma_D$ runs over two vertical angles of diapason $\phi < \pi$. In fact, we have $\phi = \pi - \alpha$ where α is the angular length of the circle through x and the points ± 1. The diapason is almost complete for p close to the diameter of D_+ and is very small for p close to the arc ∂D_+, see Fig. 7.2.

The inversion problem for the pencil Σ is reduced to the Radon transform in the plane in the following three steps. Choose coordinates (x, y) in $\mathbf{E}^2$ in such a way that $\mathbf{E}_+ = \{y \geq 0\}$, and $\mathbf{D}$ is the unit disc with centre at the origin.

Step 1. Introduce the complex coordinate $z = x + \imath y$ in $\mathbf{D}$ and apply the transform

$$z_1 = F(z) \doteq \frac{\imath - z}{\imath + z}.$$

The image of $\mathbf{E}_+$ is the unit disc D_K and the image of $D_+ = \mathbf{E}_+ \cap \mathbf{D}$ is the right half-disc. Any arc $A \in \Sigma$ is transformed to the circular arc $F(A)$ in D_K that is orthogonal to the boundary. This transform is conformal hence factorable.

Step 2. Apply the mapping

$$w_1 = G(z_1) = \frac{2z_1}{1 + |z_1|^2}.$$

It maps D_K onto the identical disc D_B and the right half of D_K to the right half of D_B. It is factorable for the family of arcs $\{F(A)\}$, see Example 2, Chapter 3. The image of an arc $F(A)$ is the chord with the same ends.

Step 3. Apply the projective transform

$$(u, v) = P(u_1, v_1) \quad u = \frac{1}{u_1}, \; v = \frac{v_1}{u_1}.$$

Figure 7.2: Arcs in half-disc and its images by transformations F, GF and PGF

The vertical diameter of D_B maps to the improper projective line and the unit circle is transformed to the hyperbola $v_1^2 + 1 = u_1^2$. The image of the disc D_B is equal to the set $W = \{v_1^2 + 1 < u_1^2\}$; the image of the right half is equal to the right connected component U of W. The image of a chord $L \subset D_B$ is a chord in the set U with ends in the hyperbola. By the transitivity property, this mapping is factorable too. Take the composition $Q = PGF$; it follows from the previous formulae that it is factorable with the Jacobian factors

$$j(z) = \frac{4y}{(1-|z|^2)^2}, \ J(A) = \left(1 + \left(\frac{ab-1}{a-b}\right)^2\right)^{1/2},$$

where a, b are the ends of A. This implies the formula

$$\int_{Q(A)} \phi \, \mathrm{d}s^* = J(A) \int_A f \, \mathrm{d}s, \tag{7.2}$$

where $\phi \doteq j^{-1} f$ and $\mathrm{d}s^*$ is the Euclidean line element in U.

The support of the function ϕ is a compact subset of U and curve $Q(A)$ is an arbitrary finite chord of the hyperbola ∂U. Let ψ be the angle of the normal to $Q(A)$; we have $|\psi| < \pi/4$. Vice versa, an arbitrary line in U whose normal has angle in this diapason, is a finite chord.

Remark. There are three classical models of the Lobachewski plane:

(i) Poincare's model in the half-plane $\mathbf{E}_+$ with the metric $y^{-2}\mathrm{d}s^2$, geodesics are the arcs $A \in \Sigma$;

(ii) Klein's model D_K in the disk $\mathbf{D}$ with the metric $\left(1-|z|^2\right)^{-2}\mathrm{d}s^2$; geodesics are arcs orthogonal to ∂D, and

(iii) Beltrami's model D_B in $\mathbf{D}$; the metric is

$$g = \left(1-x^2-y^2\right)^{-2}\left[\left(1-y^2\right)\mathrm{d}x^2 + 2xy\mathrm{d}x\mathrm{d}y + \left(1-x^2\right)\mathrm{d}y^2\right].$$

The geodesics are chords. The mappings $F : E_+ \to D_K$, $G : D_K \to D_B$ are isometries between these models. These mappings are not, of course, isometries for Euclidean metrics in $\mathbf{E}_+$ and $\mathbf{D}$, but they are factorable for the families of hyperbolic geodesics.

Corollary 7.1 *For functions with compact support in the half-disc $\mathbf{D}_+$, the integral transform on the pencil Σ_D is reduced to the Radon transform with data in the limited angle diapason $|\psi| < \pi/4$.*

From (7.2) we know the integral of the function

$$\phi(u,v) = (4y)^{-1}(1-x^2-y^2)^2 f(x,y)$$

along an arbitrary proper chord $L = Q(A)$ against the Euclidean line element $\mathrm{d}s^*$. This is a continuous function with compact support in W. By the slice theorem we have for any $-\pi/4 < \theta < \pi/4$ and any $t \in \mathbb{R}$,

$$\hat{\phi}(t\cos\theta, t\sin\theta) = \int \exp(-\mathrm{j}qt)\int_{L(q,\theta)} \phi \mathrm{d}s^*,$$

where we specify $L = L(q,\theta) = \{(u,v);\ \cos\theta\, u + \sin\theta\, v = q\}$. Denote $A(q,\theta) \doteq Q^{-1}(L(q,\theta))$ and we have $J(A(q,\theta)) = \left(q^2-\cos 2\theta\right)_+^{-1/2}$. Then by (7.2)

$$\int_{L(q,\theta)} \phi \mathrm{d}s^* = \left(q^2-\cos 2\theta\right)_+^{-1/2} Mf(A(q,\theta)),$$

which yields

$$\hat{\phi}(t\cos\theta, t\sin\theta) = \int_{\cos 2\theta}^{\infty} \exp(-\mathrm{j}qt)\frac{Mf(A(q,\theta))}{\sqrt{q^2-\cos 2\theta}}\mathrm{d}q,\ |\theta| < \frac{\pi}{4}. \tag{7.3}$$

Thus the Fourier transform $\hat{\phi}(\sigma,\tau)$ of ϕ is known in the cone $K \doteq \{(\sigma,\tau) : \sigma^2 \geq \tau^2\}$.

Remark. The right side of (7.3) depends on the integral of f along the arcs $A(q,\theta)$ with a constant angle θ; the quantity $x_A \doteq -\cot\theta = (1+ab)/(a+b)$ is constant. Consider the complexification of the plane $\mathbf{E}^2$. An arbitrary circle A is the real part of a complex conic $A_{\mathbb{C}}$ that contains the points $p_A^{\pm} \doteq (x_A, y_A)$ with the ordinates $y_A = \sqrt{1-x_A^2} = \sqrt{-\cos 2\theta}\csc\theta$. Consequently the integral in (7.3) is taken over the curve of arcs A, whose complexification $A_{\mathbb{C}}$ passes through the points $p_A^{\pm}$.

Now we use the interpolation method of Section 1.5 to reconstruct this function outside K:

$$\psi(\sigma) = \exp\left(\pi\sqrt{\delta^2-\sigma^2}\right)\int_{\Gamma}\frac{\sin\left(\pi\sqrt{\lambda^2-\delta^2}\right)}{\pi|\lambda-\sigma|}\psi(\lambda)\mathrm{d}\lambda, \quad \operatorname{Re}\sqrt{\delta^2-\sigma^2}>0, \quad (7.4)$$

where $\sigma \notin \Gamma \doteq (-\infty,-\delta)\cup(\delta,\infty)$ and δ is an arbitrary positive number. The formula (7.4) is valid for an arbitrary function $\psi \in L_2(\mathbb{R})$ such that $\operatorname{supp}\hat{\psi} \subset [-1,1]$. The support of the function ϕ is compact and hence is contained in a strip $|u-a| \le r$. Apply the interpolation method to the function $\psi_\tau(\sigma) \doteq \hat{\phi}_1(\sigma,\tau)$ taking τ as a parameter and $\phi_1(u,v) \doteq \phi(ru+a,v)$. We have $\hat{\phi}_1(\sigma,\tau) = \exp(\mathrm{j}ar^{-1}\sigma)\hat{\phi}(r^{-1}\sigma,\tau)$. The right side is known for $|\sigma| > r|\tau|$. We set $\delta = r|\tau|$ and get for an arbitrary τ the equation

$$\hat{\phi}(\sigma,\tau) = \exp\left(\pi r\sqrt{\tau^2-\sigma^2}\right)\int_{\lambda^2\ge\tau^2}\frac{\sin\left(\pi r\sqrt{\lambda^2-\tau^2}\right)\exp\left(\mathrm{j}a(\lambda-\sigma)\right)}{\pi|\lambda-\sigma|}\hat{\phi}(\lambda,\tau)\mathrm{d}\lambda \quad (7.5)$$

for an arbitrary σ,τ. Now we apply the inverse Fourier transform and recover the function f.

Theorem 7.2 *For an arbitrary function $f \in L_2(\mathbf{E}^2)$ such that* $\operatorname{supp} f \subset \mathbf{D}_+$ *the formulae*

$$f(x,y) = \frac{y}{\pi^2(1-x^2-y^2)^2}\int_2 \exp(\mathrm{j}(u(x,y)\sigma + v(x,y)\tau))\hat{\phi}(\sigma,\tau)\mathrm{d}\sigma\mathrm{d}\tau,$$

(7.3), and (7.5) give a reconstruction from the data $Mf(A)$, $A \in \Sigma_D$.

7.3 Hemispherical integrals in space

Denote by $\Sigma(\mathbf{H})$ the variety of all spheres and hyperplanes in a Euclidean space $\mathbf{E}^n$ that are orthogonal to the hyperplane $\mathbf{H} \doteq \{x_n = 0\}$. By $S(a,r) \in \Sigma$ denote a sphere with the centre $a \in \mathbf{H}$ and the radius $r > 0$. For a continuous function f, $\operatorname{supp} f \subset \mathbf{E}_+ \doteq \{x_n \ge 0\}$ define the spherical integral transform M, see (7.1). Any odd function f with respect to x_n has zero spherical integrals. On the other hand the completeness condition is fulfilled for any point x in the open half-plane $\{x_n > 0\}$. Therefore we can expect that any even function can be reconstructed. This is the case and, moreover, there are several reconstruction formulae.

Method of harmonic analysis. The following reconstruction method is close to that of Fawcett [23]. Let J_ν be the Bessel function of order ν.

Theorem 7.3 *Any real function $f \in L_2(\mathbf{E}^n)$ that is even in x_n and decreases sufficiently fast at infinity, can be reconstructed from spherical integrals $g(a,r) = Mf(a,r)$, $a \in \mathbf{H}$, $r > 0$ by the formula*

$$f(x', x_n) = \pi (2\pi)^{-n/2} \int_{\mathbf{E}^*} \exp(\imath(\langle \xi, x' \rangle + \eta x_n)) |\eta| \, \tilde{g}\left(\xi, \left(|\xi|^2 + \eta^2\right)^{1/2}\right) \mathrm{d}\xi \mathrm{d}\eta, \tag{7.6}$$

where

$$\tilde{g}(\theta, \rho) \doteq \rho^{n/2-1} \int_0^\infty \int_{\mathbf{H}} \exp(-\imath \langle \theta, a \rangle) J_{n/2-1}(\rho r) r^{1-n/2} Mf(a,r) \,\mathrm{d}a\mathrm{d}r.$$

◀ Write Mf as an integral over the unit sphere against the area element $\mathrm{d}\omega$ and apply the equation $f = F^* \hat{f}$:

$$\begin{aligned}
\hat{g}(\theta, \rho) &= \rho^{n/2-1} \int_{\mathbf{H}} \exp(-\imath \langle \theta, a \rangle) \,\mathrm{d}a \int_{\mathbb{R}} J_{n/2-1}(\rho r) r^{1-n/2} Mf(a,r) \,\mathrm{d}r \\
&= (2\pi)^{-n} \rho^{n/2-1} \int_{\mathbf{H}} \exp(-\imath \langle \theta, a \rangle) \,\mathrm{d}a \int_0^\infty J_{n/2-1}(\rho r) r^{n/2} \mathrm{d}r \\
&\quad \times \int_{y=a+r\omega} \mathrm{d}\omega \int_{\mathbf{E}^*} \exp(\imath \langle \xi, y' \rangle + \imath \eta y_n) \hat{f}(\zeta, \eta) \,\mathrm{d}\zeta \mathrm{d}\eta.
\end{aligned}$$

By changing the variables $z = (y' - a, y_n)$ and the order of integration, we obtain the inner integral

$$\int_{|z|=r} \exp(\imath \langle \xi, y' - a \rangle + \imath \eta y_n) \,\mathrm{d}\omega = (2\pi)^{n/2} (r\sigma)^{1-n/2} J_{n/2-1}(r\sigma),$$
$$\sigma = \left(|\xi|^2 + \eta^2\right)^{1/2}.$$

From the inversion formula for the Fourier-Hankel transform

$$\int_0^\infty J_{n/2-1}(r\rho) J_{n/2-1}(r\sigma) r dr = \rho^{-1} \delta(\sigma - \rho).$$

This simplifies the inner integral in the right side to

$$\int_0^\infty J_{n/2-1}(\rho r) r^{n/2} (r\sigma)^{1-n/2} J_{n/2-1}(r\sigma) \,\mathrm{d}r = \rho^{-n/2} \delta(\sigma - \rho).$$

Taking into account that

$$\int_{\mathbf{H}} \exp(-\imath \langle \theta - \xi, a \rangle) \,\mathrm{d}a = (2\pi)^{n-1} \delta(\theta - \xi),$$
$$\rho^{-1} \delta(\sigma - \rho) \,\mathrm{d}\xi \mathrm{d}\eta = |\eta|^{-1} \delta(\sigma - \rho) \,\mathrm{d}\xi \mathrm{d}\sigma,$$

yields

$$\begin{aligned}\tilde{g}(\theta,\rho) &= (2\pi)^{n/2-1}\int_{\mathbf{H}}\int_{\mathbb{R}}\delta(\theta-\xi)\,\rho^{-1}\,\delta(\rho-\sigma)\,\hat{f}(\xi,\eta)\,\mathrm{d}\xi\mathrm{d}\eta \\ &= (2\pi)^{n/2-1}\int\delta(\theta-\xi)\,|\eta|^{-1}\,\delta(\rho-\sigma)\,\hat{f}(\theta,\eta)\,\mathrm{d}\xi\mathrm{d}\sigma \\ &= (2\pi)^{n/2-1}\,|\eta|^{-1}\left[\hat{f}\left(\theta,\left(\rho^2-|\theta|^2\right)^{1/2}\right)+\hat{f}\left(\theta,-\left(\rho^2-|\theta|^2\right)^{1/2}\right)\right] \\ &= 2\,(2\pi)^{n/2-1}\,|\eta|^{-1}\,\hat{f}\left(\theta,\left(\rho^2-|\theta|^2\right)^{1/2}\right),\end{aligned}$$

since the function $\hat{f}(\xi,\eta)$ is even in η. ▶

Corollary 7.4 *The condition* $\tilde{g}(\theta,\rho)=0$ *for* $\rho<|\theta|$ *is necessary for* g *to be the spherical transform of a function.*

Backprojection method. See [23]. Introduce the backprojection operator defined for functions $g=g(a,r)$:

$$M^{\sharp}g(x)=\int\int r^{1-n}g(a,r)\,\delta(r-|x-a|)\,\mathrm{d}x\mathrm{d}r.$$

Denote by H_n the Hilbert operator in the variable x_n.

Theorem 7.5 ([23]) *Under the same conditions the function* f *can be recovered for even* n *by the formula*

$$f(x)=-\frac{\pi}{\mathrm{j}^n}H_n\frac{\partial}{\partial x_n}\Delta^{n/2-1}M^{\sharp}g(x)$$

and for odd n *by*

$$f(x)=-\frac{\pi}{s_{n-1}\mathrm{j}^{n+1}}H_n\frac{\partial}{\partial x_n}\Delta^{(n-1)/2}\int\frac{M^{\sharp}g(y)\,\mathrm{d}y}{|x-y|^{n-1}},$$

where $s_{n-1}=\pi^{(n-1)/2}/\Gamma((n-1)/2)$.

Factorization method. We can reduce the problem to the Radon transform by applying an appropriate factorable mapping. Take another Euclidean space $\mathbf{F}$ of the same dimension with coordinates $y=(y_1,\dots,y_n)$ and consider the open unit ball $\mathbf{B}$ in $\mathbf{F}$. Define the mapping $\mathsf{B}:\mathbf{E}_+\to\mathbf{B}$ by

$$\mathsf{B}:x\mapsto y=\left(\frac{2x_1}{1+x^2},\dots,\frac{2x_{n-1}}{1+x^2},\frac{1-x^2}{1+x^2}\right),\ x^2=|x|^2.$$

Denote by $L(p,\omega)$ the hyperplane $\{\langle\omega,y\rangle=p\}$ in $\mathbf{F}$ where $|\omega|=1$, $p\in\mathbb{R}$. It has non-empty intersection with $\mathbf{B}$ if and only if $|p|<1$.

Proposition 7.6 *For any sphere $S \subset \mathbf{E}^n$ orthogonal to $\mathbf{H} = \partial \mathbf{E}_+$ the image $L = \mathsf{B}(S \cap \mathbf{E}_+)$ is a hyperplane in $\mathbf{B}$. The mapping B is factorable for the family $\Sigma(\mathbf{H})$ of spheres orthogonal to $\mathbf{H}$ with the Jacobian factors*

$$j_{\mathsf{B}}(x) = \frac{2^n x_n}{\left(1+x^2\right)^n}, \; J_{\mathsf{B}}(S) = \left(1-p^2\right)^{-1/2}, \tag{7.7}$$

where $\mathsf{B}(S) = L(p, \omega)$.

◀ The equation $\langle \omega, y \rangle = p$ is equivalent to

$$\sum_1^{n-1} \left(x_i - \frac{\omega_i}{p+\omega_n}\right)^2 + x_n^2 = \frac{1-p^2}{\left(p+\omega_n\right)^2},$$

which yields

$$\mathsf{B}(S(a,r)) = L(p,\omega), \; \omega = (\omega', \omega_n), \; a = \frac{\omega'}{p+\omega_n}, \; r = \frac{\left(1-p^2\right)^{1/2}}{|\omega_n + p|}, \tag{7.8}$$

where the vector (p, ω) is defined up to sign from data (a, r). If $p \to -\omega_n$, the sphere $S(a,r)$ tends to a hyperplane with the normal vector $(\omega', 0)$ orthogonal to $\mathbf{H}$. The inverse mapping is given by

$$\mathsf{B}^{-1}: \; y \mapsto x = (x', x_n), \; x' = \frac{y'}{1+y_n}, x_n = \frac{\sqrt{1-y^2}}{1+y_n}.$$

The half-space $\mathbf{E}_+$ with the metric $\mathrm{d}\sigma_{\mathbf{E}}^2 \doteq x_n^{-2} \mathrm{d}s_{\mathbf{E}}^2$ is the Poincaré model of hyperbolic space of constant curvature. On the other hand, the ball $\mathbf{B}$ with the metric

$$\mathrm{d}\sigma_{\mathbf{F}} \doteq \left(1-y^2\right)^{-1} \sum \mathrm{d}y_i^2 + \left(1-y^2\right)^{-2} \left(\sum y_i \mathrm{d}y_i\right)^2$$

is the Beltrami-Klein model of a hyperbolic space. The mapping B is the isometry with respect to these metrics, see [98]. It follows that

$$\begin{aligned} \left(\mathsf{B}^{-1}\right)^* \left(\mathrm{d}s_{\mathbf{E}}^2\right) &= x_n^2 \mathrm{d}\sigma_{\mathbf{E}}^2 = x_n^2 \mathrm{d}\sigma_{\mathbf{F}}^2 \\ &= \left(1+y_n\right)^{-2} \left[\sum \mathrm{d}y_i^2 + \left(1-y^2\right)^{-1} \left(\sum y_i \mathrm{d}y_i\right)^2\right], \end{aligned}$$

which means that this metric is equal to

$$\begin{aligned} \sum \mathrm{d}x_i^2 &= \sum g^{ij} \mathrm{d}y_i \mathrm{d}y_j, \; g^{ij} = \left(1+y_n\right)^{-2} \left[\delta_j^i + \left(1-y^2\right)^{-1} y_i y_j\right] \\ \sum \mathrm{d}y_i^2 &= \sum g_{ij} \mathrm{d}x_i \mathrm{d}x_j, \; g_{ij} = \left(1+y_n\right)^2 \left[\delta_j^i - y_i y_j\right]. \end{aligned}$$

It follows that

$$\mathrm{d}y = \left(1-y^2\right)^{1/2} \left(1+y_n\right)^n \mathrm{d}x.$$

The equation $\langle \omega, y\rangle - p = 0$ defines the hyperplane $L(\omega, p)$ in $\mathbf{F}$ and the sphere $S(a, r)$ in $\mathbf{E}^n$. Therefore

$$\frac{\mathrm{d}S_y(L(\omega,p))}{\mathrm{d}S_x(S(a,r))} = \frac{|\nabla_y \langle \omega, y\rangle|}{|\nabla_x \langle \omega, y\rangle|}\frac{\mathrm{d}y}{\mathrm{d}x},$$

where $|\nabla_y \langle \omega, y\rangle| = 1$ and

$$|\nabla_x \langle \omega, y\rangle|^2 = \sum g_{ij}\omega_i\omega_j = (1+y_n)^2\left[\sum \omega_i^2 - \langle \omega, y\rangle^2\right] = (1+y_n)^2\left(1-p^2\right).$$

Finally

$$\frac{\mathrm{d}S_y(L(\omega,p))}{\mathrm{d}S_x(S(a,r))} = \left(1-y^2\right)^{1/2}(1+y_n)^{n-1}\left(1-p^2\right)^{-1/2} = j_{\mathsf{B}}(x)\, J_{\mathsf{B}}(a,r),$$

which agrees with (7.7). ▶

From (7.8) we derive p and ω as functions of a and r:

$$p = \frac{1+|a|^2 - r^2}{g(a,r)}, \quad J_{\mathsf{B}}(S(a,r)) = \frac{g(a,r)}{2r},$$

$$\omega_n = \frac{1-|a|^2+r^2}{g(a,r)}, \quad \omega' = \frac{2a}{g(a,r)},$$

where $g(a,r) = \sqrt{\left(1+(|a|+r)^2\right)\left(1+(|a|-r)^2\right)}$ is the geometric mean of two intervals of extremal length joining the point $(0,\ldots,0,1)$ with points of $n-2$-sphere $S(a,r)\cap\mathbf{H}$.

Corollary 7.7 *For a function f with compact support in $\mathbf{E}_+$ we have*

$$R\phi(\omega,p) = J_{\mathsf{B}}(a,r)\, Mf(a,r),$$

where $\phi(\mathsf{B}(x)) \doteq j_{\mathsf{B}}^{-1}(x)\, f(x)$ and R means the Radon transform in $\mathbf{F}$.

Corollary 7.8 *The spherical integral transform defined on the pencil $\Sigma(\mathbf{H})$ of spheres in $\mathbf{E}$ orthogonal to $\mathbf{H}$ can be inverted for functions with support in $\mathbf{E}_+$ by the formula*

$$f = j_{\mathsf{B}} R^\sharp\left(-\frac{\mathbb{H}}{2\pi}\frac{\partial}{\partial p}\right)^{n-1} J_{\mathsf{B}} Mf,$$

where $J_{\mathsf{B}}Mf$ is written in coordinates (p,ω). Any other inversion formula for the Radon transform R can be applied.

Remark. The Plancherel theorem and range conditions can be translated for the Funk transform on the family of spheres $\Sigma(\mathbf{H})$ by the same reduction.

7.4 Incomplete data

Let $\mathbf{B}$ be the unit ball in $\mathbf{E}^n$, denote by $\mathbf{B}_+ = \mathbf{B} \cap \mathbf{E}_+$ the half-ball. For functions f supported in $\mathbf{B}_+$ there is one method of reduction to the Radon transform. Take another Euclidean space $\mathbf{F}$ of dimension n and consider the interior Q of the one-fold hyperboloid $\{q(z) \geq 1,\ z_n > 0\} \subset \mathbf{F}$, where

$$q(z) = z_n^2 - z_1^2 - \cdots - z_{n-1}^2.$$

Consider the mapping Q: $\mathbf{B}_+ \to \mathbf{F}$ given by

$$x \mapsto z = \left(\frac{2x_1}{1 - x^2}, \ldots, \frac{2x_{n-1}}{1 - x^2}, \frac{1 + x^2}{1 - x^2} \right).$$

The mapping Q is an invertible mapping from $\mathbf{B}_+$ onto Q. If $S = S\,(a, r)$ is a sphere in $\mathbf{E}$ which is orthogonal to $\mathbf{H} = \partial \mathbf{E}_+$, then $L \doteq \mathsf{Q}\,(S)$ is a hyperplane in $\mathbf{F}$. We have $L = L\,(\omega, p)$ where

$$a = \frac{\omega'}{p + \omega_n}, \quad r = \frac{\sqrt{p^2 - q\,(\omega)}}{p + \omega_n} \tag{7.9}$$

and we assume that $p + \omega_n > 0$.

Proposition 7.9 *The mapping Q is factorable for the family of spheres S orthogonal to $\mathbf{H}$ with the factors*

$$j_{\mathsf{Q}}\,(x) = \frac{2^n x_n}{(1 - x^2)^n}, \quad J_{\mathsf{Q}}\,(S) = \left(p^2 - q\,(\omega)\right)^{-1/2}, \tag{7.10}$$

where $\mathsf{Q}\,(S) = L\,(\omega, p)$.

◀ The map Q equals the composition of the map B introduced in the previous section and the projective mapping $P : y \mapsto z = (y_1/y_n, \ldots, y_{n-1}/y_n, 1/y_n)$ and P is factorable for the family of hyperplanes $L = \mathsf{B}\,(S)$, $S \in \Sigma\,(\mathbf{H})$. We apply Proposition 3.1 of Chapter 3. ▶

It follows that for any function f supported by $\mathbf{E}_+$ and the function $f^*\,(z) = \phi\left(\mathsf{Q}^{-1}\,(z)\right), \phi = j_{\mathsf{Q}}^{-1} f$, we can reduce the spherical transform to the Radon transform as follows:

$$R\phi\,(\omega, p) = \left(p^2 - q\,(\omega)\right)_+^{-1/2} Mf\,(a, r). \tag{7.11}$$

Proposition 7.10 *If $S \Subset \mathbf{B}_+$, then the intersection of $L\,(\omega, p) = \mathsf{Q}\,(S)$ with Q is compact, $p > q\,(\omega)$ and vice versa.*

◀ The inclusion $S\,(a, r) \Subset \mathbf{B}_+$ means that $|a| + r \leq 1$. By (7.9) this is equivalent to the inequality

$$|\omega'| + \sqrt{p^2 - \omega_n^2 + |\omega'|^2} < p + \omega_n.$$

This implies that $|\omega'| < \omega_n$ and vice versa. The last inequality can be written in the form $q\,(\omega) > 0$ which means that the intersection $L\,(\omega, p) \cap Q$ is compact. ▶

Corollary 7.11 *If the data of spherical integrals $Mf(S)$ is available only for $S \in \mathbf{B}_+$, then we know from* (7.11) *data of the Radon transform of ϕ on all hyperplanes $L(\omega, p)$ such that $q(\omega) > 0$.*

We can interpolate the Radon data for all $\omega \in \mathbf{S}^{n-1}$ by the method of Section 5.3 and apply any inversion formula.

7.5 Spheres centred on a sphere

Let B be a ball in $\mathbf{E}^n$ of radius b and $S = \partial B$. Consider the variety $\Sigma(S)$ of spheres $S(a, r)$ with centres $a \in S$. We shall reconstruct a function f defined in $\mathbf{E}^n$ from the data $Mf(S) = Mf(a, r)$ for $S = S(a, r) \in \Sigma(S)$. The reconstruction is much easier if n is odd. Given a density σ on S, we define for a function $g = g(p, r)$ in $S \times \mathbb{R}_+$, the spherical backprojection to be a function in $\mathbf{E}^n$,

$$M^\sharp g(x) = \int_S g(p, |x - p|)\,\sigma,$$

where σ is the Euclidean area density.

Theorem 7.12 ([25]) *If $n \geq 3$ is odd, then for any continuous function f and point $x \in B$ we have*

$$f(x) = \frac{(-1)^{(n-1)/2}}{8\pi^{n-1}b}\Delta_x M^\sharp\left[\left(\frac{\partial}{\partial r^2}\right)^{n-3}\left(\frac{1}{r}Mf\right)\right] \tag{7.12}$$

and

$$f(x) = \frac{(-1)^{(n-1)/2}}{8\pi^{n-1}b}\Delta_x M^\sharp\left[\frac{1}{r}\left(\frac{\partial}{\partial r^2}\right)^{n-3} Mf\right]. \tag{7.13}$$

◄ Set $m = n - 3,\ s = r^2,\ \partial_s = \partial/\partial s$ and choose the origin in the centre of S. Write

$$M^\sharp\left[\partial_s^m\left(\frac{1}{r}Mf\right)\right](x) = \int_S\int_0^\infty\int_B \partial_s^m\frac{\delta(|y-p|-r)}{r}f(y)\,\mathrm{d}y\,\delta(|x-p|-r)\,\mathrm{d}r\sigma(p).$$

We have

$$r^{-1}\delta(|y-p|-r) = 2\delta\left(|y-p|^2 - s\right), \partial_s^m\frac{\delta(|y-p|-r)}{r} = 2\delta^{(m)}\left(|y-p|^2 - s\right),$$

since the integer m is odd, and integrate

$$\begin{aligned}\int \delta^{(m)}\left(|y-p|^2 - s\right)\delta(|x-p|-r)\,\mathrm{d}r &= \delta^{(m)}\left(|x-p|^2 - |y-p|^2\right) \\ &= \delta^{(m)}\left(|x|^2 - |y|^2 + 2\langle y - x, p\rangle\right).\end{aligned}$$

The right side is the pull back of the generalized function $\delta^{(m)}(t)$ by the mapping $(y,p) \mapsto t = |x|^2 - |y|^2 + 2\langle y-x, p\rangle$ (see the definition of pull back in Section 1.8). This yields

$$M^{\sharp}\left[\partial_s^m \frac{1}{r} Mf\right](x) = 2\int_B \left(\int_S \delta^{(m)}\left(|x|^2 - |y|^2 + 2\langle y-x,p\rangle\right)\sigma(p)\right) f(y)\,\mathrm{d}y.$$

Now the argument of the Dirac function is linear in p and

$$\delta^{(m)}\left(|x|^2 - |y|^2 + 2\langle y-x,p\rangle\right) = \frac{1}{(2|y-x|)^{m+1}}\delta^{(m)}(\langle \omega, p\rangle - z), \tag{7.14}$$

$$\text{where } z = \frac{|y|^2 - |x|^2}{2|y-x|},\ \omega = \frac{y-x}{|y-x|},$$

which simplifies the integral over S:

$$\int_S \delta^{(m)}\left(|x|^2 - |y|^2 + 2\langle y-x,p\rangle\right)\sigma = \frac{1}{(2|y-x|)^{n-2}}\partial_z^m \int_S \delta(\langle \omega, p\rangle - z)\,\sigma.$$

The integral on the right side is equal to $2\pi^{(n-1)/2}\Gamma((n-1)/2)^{-1} b\left(b^2 - z^2\right)^{m/2}$ for $|z| < b$. It follows that the derivative

$$\partial_z^m \int_S \delta(\langle \omega, p\rangle - z)\,\sigma(p) = (-1)^k m! \frac{2\pi^{(n-1)/2} b}{\Gamma((n-1)/2)}$$

is constant for $|z| \le b$. This yields

$$M^{\sharp}\left[\partial_s^m \frac{1}{r} Mf\right](x) = (-1)^k \frac{m!\,\pi^{(n-1)/2} b}{2^{n-4}\Gamma((n-1)/2)} \int \frac{f(y)\,dy}{|y-x|^{n-2}}.$$

The singular kernel is a fundamental solution for the Laplace operator:

$$\Delta_x |x-y|^{2-n} = -\frac{4\pi^{n/2}}{\Gamma(n/2-1)}\delta(x-y).$$

This implies

$$\begin{aligned}\Delta M^{\sharp}\left[\partial_s^m \frac{1}{r} M\right] &= (-1)^{k+1} \frac{m!\,\pi^{(n-1)/2} b}{2^{n-6}\Gamma((n-1)/2)} \frac{\pi^{n/2}}{\Gamma(n/2-1)}\delta(x-y) \\ &= (-1)^{k+1} 2^4 \pi^{n-1} b\,\delta(x-y),\end{aligned}$$

since $\Gamma((n-1)/2)\Gamma(n/2-1) = \pi^{1/2}2^{3-n}m!$, which yields (7.12). To check the second formula, we write

$$\begin{aligned}M^{\sharp}\left[\frac{1}{r}\partial_s^m Mf\right] &= \int_S \int_0^\infty \int_B \frac{\delta(|x-p|-r)}{r}\partial_s^m \delta(|y-p|-r) f(y)\,\mathrm{d}y\,\mathrm{d}r\,\sigma(p) \\ &= 2\int_S\int_B\int_0^\infty \delta\left(|x-p|^2 - s\right)\partial_s^m \delta(|y-p|-r)\,\mathrm{d}r\,\sigma(p)\,f(y)\,\mathrm{d}y.\end{aligned}$$

We have $\delta(|y-p|-r)\,\mathrm{d}r = \delta\left(|y-p|^2 - s\right)\mathrm{d}s$, which implies that the inner integral is equal to

$$\int_0^\infty \delta\left(|x-p|^2 - s\right)\partial_s^m \delta\left(|y-p|^2 - s\right)\,\mathrm{d}s = \delta^{(m)}\left(|x-p|^2 - |y-p|^2\right)$$
$$= \delta^{(m)}\left(|x|^2 - |y|^2 + 2\langle y-x, p\rangle\right).$$

We apply (7.14) and follow the previous arguments. This yields (7.13). ►

7.6 Spheres tangent to a manifold

Normalizing the integral (7.1), we obtain the spherical mean transform:

$$Nf(a,r) = s_n^{-1} r^{1-n} Mf(a,r) = \int_{|\omega|=1} f(a+r\omega)\,\mathrm{d}\omega,$$

where we use the abbreviation $s_n = 2\pi^{n/2}/\Gamma(n/2)$. For any continuous f the function Nf is extended by continuity to the space $\mathbf{E}\times\mathbb{R}_+$ where $\mathbf{E} = \mathbf{E}^n$ and $\mathbb{R}_+$ is the half-line of non-negative numbers; we have $Nf(x,0) = f(x)$. The spherical mean transform $g(x,r) \doteq Nf(x,r)$ satisfies the Darboux equation in $\mathbf{E}\times\mathbb{R}_+$:

$$Ag \doteq \left(r\frac{\partial^2}{\partial r^2} + (n-1)\frac{\partial}{\partial r} - r\Delta_x\right) g = 0.$$

The reconstruction problem is as follows: to recover a function f from data of its spherical mean transform on the variety $\Sigma \subset \mathbf{E}\times\mathbb{R}_+$. If $\dim\Sigma = n$, the data of $Nf(S)$ for $S \in \Sigma$ are not redundant. For reconstruction of the function f it is sufficient to find the solution of the Darboux equation on the hyperplane $r = 0$. The principal part of the Darboux operator is equal to the wave operator $r\Box$ where $\Box = \partial_r^2 - \Delta_x$ is the wave operator with velocity 1. A hypersurface $\Sigma \subset X$ is *characteristic* for the Darboux (and for the wave) operator at a point $(x,r) \in X, r > 0$, if the principal symbol vanishes on the conormal vector ν to Σ at x, i.e., if $\nu_r^2 - \nu_x^2 = 0$, where $\nu_x \in E$ and $\nu_r \in \mathbb{R}$ are the components of ν.

We now focus on the special case: $\Sigma = \Sigma(Y)$; for a submanifold $Y \subset \mathbf{S}$ we denote $\Sigma(Y)$ the submanifold of spheres in $\mathbf{S}$ tangent to Y.

Proposition 7.13 *The variety $\Sigma(Y)$ is characteristic at each of its points.*

It follows that the reconstruction problem can be reduced to the characteristic Cauchy problem: *to recover a solution g of the Darboux equation on the boundary $\mathbf{E}\times\{0\}$ from knowledge of its value on a characteristic hypersurface.* We have then $f(x) = g(x,0)$. We obtain in this way a reconstruction formula for an arbitrary bounded domain $D \subset \mathbf{E}$ with smooth boundary ∂D. Define the function $\rho = \rho(y) \doteq \min_{z\in\partial D}|y-z|$ in D; it is smooth everywhere in D except for curvature centres of the boundary.

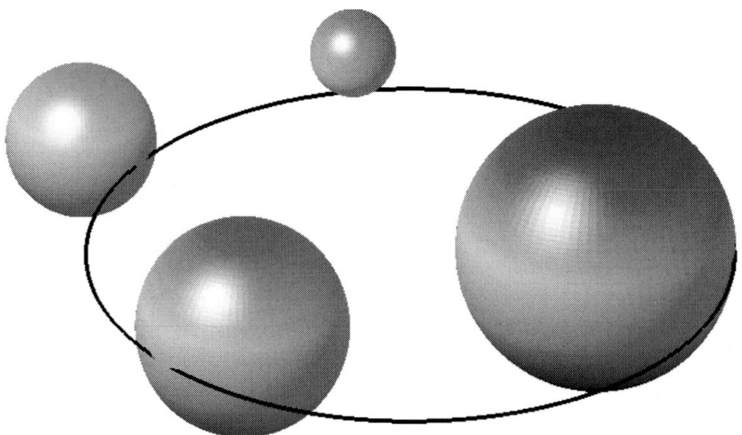

Figure 7.3: Spheres tangent to a curve

Theorem 7.14 *For arbitrary $n \geq 2$, any function f in D can be reconstructed from data of spherical transform Mf for spheres $S \subset D$ tangent to the boundary as follows:*

$$f(x) = \frac{1}{2\mathrm{j}^{n-1}} \int_D \delta^{(n-1)}(q) \, \Phi_n(x, y) \, Mf(y, r) \,|_{r=\rho} \mathrm{d}y \tag{7.15}$$

for odd n where

$$\Phi_n(x, y) = 2 \langle y - x, \nabla \rho \rangle - 2\rho + q \left(\frac{1}{\rho} + \frac{\Delta \rho}{n-1} \right),$$

and

$$f(x) = \frac{(n-1)!}{\mathrm{j}^n} \int_D \left[q^{-n} \right] \Phi_n(x, y) \, Mf(y, r) \,|_{r=\rho} \mathrm{d}y \tag{7.16}$$

for even n, where $q = \left(r^2 - |y - x|^2 \right)/2$ and the generalized function $\left[q^{-n} \right]$ is defined in Section 1.5.

Remark. The integrals (7.15) and (7.16) have sense, at least, for generic D and $x \in D$. Indeed, the wave front of $[q^{-n}]$ and of $\delta^{(n-1)}(q)$ coincides with the conormal bundle $N^*(Q_x)$ of the cone $Q_x \doteq \{q = 0\} \subset D \times \mathbb{R}_+$. The function $\rho(y)$ is smooth in D, except for the set C of points y where the sphere $S(y, \rho)$ has more than one point of contact with ∂D or y is a curvature centre of ∂D (multiple contact). The wave front $WF(G)$ of the function $G(y, r) \doteq r - \rho(y)$ is generically a conic Lagrange manifold in $T^*(D \times \mathbb{R}_+)$ (with singularities, because of double points of contact). Two conic Lagrange manifolds $N^*(Q_x)$ and $WF(G)$ have generically no common point. In this case the restriction of $[q^{-n}]$ and of $\delta^{(n-1)}(q)$ to the hypersurface $r = \rho(y)$ is well defined. Moreover, the set of points x such that $N^*(Q_x) \cap WF(G) \neq \emptyset$ has zero measure in D.

Theorem 7.15 *Let n be an odd integer and Y be an arbitrary closed submanifold of* $\mathbf{E}$, *$H(Y)$ be the convex hull of Y. Any sufficiently smooth function f can be explicitly reconstructed in* $\operatorname{int} H(Y)$ *from data of spherical means on $\Sigma(Y)$ as follows:*

$$f(x) = \int_{\Sigma(Y)\cap\Sigma(x)} K_{\mathbf{E}}(x,S)\, N_{\mathbf{E}} f(S),$$

where $K_{\mathbf{E}}(x,\cdot)$ is for any $x \in \operatorname{int} H(Y)$ a distribution with compact support in the manifold $\Sigma(Y)\cap\Sigma(x)$ of spheres tangent to Y and containing x.

The spherical mean transform satisfies the Darboux equation and our key idea is to consider the reconstruction problem as a Cauchy problem with "initial" data on the characteristic variety $\Sigma(Y)$. A solution u is uniquely defined from knowledge of u on $\Sigma(Y)$; we need not know the normal derivative of u. The explicit reconstruction is done by application of the singular forward fundamental solution for the adjoint Darboux operator supported by the future cone.

The propagator looks similar to the forward fundamental solutions for the wave equation, however the order of singularity is different. In particular, for odd n the propagator is a derivative of the Dirac function supported by the future cone, i.e., it possesses the strong Huygens property. In the case of even n it is a homogeneous function whose singularity on the cone is of integer degree, unlike the wave propagator whose degree is half-integer.

Theorem 7.16 *Let Y be an arbitrary non-empty closed submanifold of* $\mathbf{S}$. *An arbitrary sufficiently smooth function f in* $\mathbf{S}$ *can be reconstructed in* $\mathbf{S}$ *from data of spherical integrals $M_{\mathbf{S}} f$ on $\Sigma(Y)$ as follows:*

$$f(x) = \int_{\Sigma(Y)} K_{\mathbf{S}}(x,S)\, M_{\mathbf{S}} f(S),$$

where $K_{\mathbf{S}}(x,\cdot)$ is for any $x \in \mathbf{S}$ a distribution supported by $\Sigma(Y)$. For n odd, $K_{\mathbf{S}}$ is supported by the set $\Sigma(Y)\cap\Sigma(x)$.

In the particular case Y is a point, the reconstruction problem is reduced to inversion of the Radon transform in $\mathbf{E}$ by application of the geometric inversion mapping J with the centre in Y.

Let $\mathbf{S}$ be the unit sphere in $\mathbf{E}^{n+1}$; consider the inverse stereographic projection $\pi : \mathbf{E} \to \mathbf{S}$ with the centre $(1,0)$, where

$$\pi(x) = (y_0, y),\; y_0 = \frac{|x|^2 - 1}{|x|^2 + 1},\, y = \frac{2x}{|x|^2 + 1}.$$

The mapping π is conformal and $\pi^*(\mathrm{d}\sigma) = (1 - y_0)\,\mathrm{d}s$ where $\mathrm{d}\sigma, \mathrm{d}s$ are line elements in $\mathbf{S}$ respectively $\mathbf{E}$. The image $\pi(S)$ of an arbitrary sphere S in $\mathbf{E}$ is a sphere in $\mathbf{S}$, the image of a hyperplane H is a sphere in $\mathbf{S}$ through the centre

$(1,0)$ of the projection π. Therefore $M_{\mathbf{E}}f(S) = M_{\mathbf{S}}g(\pi(S))$ for any sphere or hyperplane S where g is a function in $\mathbf{S}$ and

$$f(x) = (1-y_0)^{n-1} g(\pi(x)) = \left(\frac{2}{1+|x|^2}\right)^{n-1} g(\pi(x))$$

is a function in $\mathbf{E}$. Therefore, inversion of the transform $I_{\mathbf{E}}$ is equivalent to inversion of $I_{\mathbf{S}}$, at least, for continuous functions f that fulfil the estimate $f = O\left(|x|^{1-n-\varepsilon}\right)$ at infinity. Then g is integrable on any sphere $S \subset \mathbf{S}$. Any sphere S divides $\mathbf{S}$ in two connected parts; we call each of them a ball.

◀ Proof of Theorem 7.16. We deduce this result from Theorem 7.14. For a subset $G \subset \mathbf{S}$ and a point $p \in \mathbf{S}\backslash G$ we denote by $H_p(G)$ the convex hull of the set $\pi_p^{-1}(G)$ in the vector space $\mathbf{E}$ where $\pi_p : \mathbf{E} \to \mathbf{S}\backslash\{p\}$ is the stereographic projection with the centre p. Now we consider three cases:

The first case: $\dim Y = n-1$. The hypersurface Y divides $\mathbf{S}$ in m connected components for some $m \geq 2$. Take a centre z in one of the components. Then f can be reconstructed in any other component by means of Theorem 7.14.

The second case: Y is a k-sphere for some $k, 0 \leq k \leq n-1$. In the case $k=0$ we apply inversion with the centre in a point $y \in Y$ and reduce the problem to inversion of the Radon transform. The case $k \geq 1$ can be done by easy arguments.

The third case: Y is not a sphere and $\dim Y < n-1$. Then for almost all points $p \in \mathbf{S}^n$, Y is contained in no $n-1$-sphere through p. Applying the inversion with the centre p, we may assume that Y is contained in no hyperplane.

Lemma 7.17 *For any point $z \in \mathbf{S}\backslash Y$ there exists a point $p \in \mathbf{S}$ such that $z \in H_p(Y)$.*

By Theorem 7.14 the function f can be reconstructed in $H_p(Y)$ and therefore at z. This completes the proof of Theorem 7.16. ▶

◀ Proof of the lemma. First take a centre p such that the convex hull $K \doteq H_p(Y)$ is not contained in a hyperplane, i.e., K is an n-body. For a point $b \in \partial K$ we denote by C_b the open asymptotic cone of the convex body K. It is easy to see that if a point $q \in \mathbf{S}\backslash K$ approaches b, the set $H_q(Y)$ tends to C_b, hence the reconstruction is possible in C_b. Then we note that for any convex n-body K we have

$$\cup_{b \in \partial K} C_b = \mathbf{E}.$$

This implies a reconstruction in $\mathbf{E} \simeq \mathbf{S}\backslash\{p\}$. Taking another centre p' we get a reconstruction at the point p. ▶

7.7 Characteristic Cauchy problem

To prove Theorem 7.14, we consider the Cauchy problem for a Darboux operator with initial data on the characteristic variety $\Sigma(Y)$. To clarify the idea, let us consider the general linear differential equation

$$A(y,D)\,u=0 \tag{7.17}$$

with smooth coefficients in an open set U in a Euclidean space $\mathbf{E}^n$.

Proposition 7.18 *Let K be compact in U with Lipschitz boundary ∂K, Γ be an open subset of ∂K and*

(i) *F_x be a distribution that fulfils the equation $A^* F_x = \delta_x$ for a point $x \in \operatorname{int} K$,*
(ii) $\operatorname{supp} F_x \cap \partial K \Subset \Gamma$,
(iii) *the trace $F_x|\Gamma$ be well defined as a generalized function in Γ and*
(iv) *Γ be a characteristic hypersurface for A. Then an arbitrary sufficiently smooth solution u of (7.17) can be reconstructed in x from data of $u|\Gamma$.*

◄ Take a smooth function ϕ in $\mathbf{E}$ such that $\phi \geq 0$ in K, $\phi < 0$ in $\mathbb{R}^n \backslash K$ and $\mathrm{d}\phi \neq 0$ on Γ. The composition $\Phi = \theta(\phi)$ is the indicator function of U where $\theta(t) = 1$ for $t \geq 0$ and $\theta(t) = 0$ otherwise. By (i) we have

$$u(x) = \delta_x(\Phi u)\,\mathrm{d}y = \int_U A^*(F_x)\,\Phi u \mathrm{d}y = \int_U A(\Phi u)\,F_x \mathrm{d}y \tag{7.18}$$

and $A = a_2 + a_1 + a_0$ where a_j is a homogeneous differential operator of degree j. By the Leibniz formula

$$A(y,D)(\Phi u) = \Phi A(y,D)\,u + B(y,\nabla\Phi,\nabla u) + u\,[a_2(y,D) + a_1(y,D)]\,\Phi,$$

where $2B(y,\xi,\eta) \doteq a_2(y,\xi+\eta) - a_2(y,\xi-\eta)$ is the symmetric bilinear form in ξ and η such that $B(y,\xi,\xi) = 2a_2(y,\xi)$. The first term vanishes by virtue of (7.17), the second and the third terms are supported in ∂K. We have $\nabla\Phi = \nabla\phi\,\delta_0(\phi)$ where δ_0 is the Dirac function of one variable and $B(y,\nabla\Phi,\nabla u) = B(y,\nabla\phi,\nabla u)\,\delta_0(\phi)$. The operator $b \doteq B(y,\nabla\phi,\nabla\cdot)$ is a tangent field such that $b\phi = B(y,\nabla\phi,\nabla\phi) = 2a_2(y,\nabla\phi)$. The right side vanishes in Γ due to (iv). This means that the field b is tangent to Γ. Any integral curve of b is a bicharacteristic of A. We have $a_1\Phi = a_1(y,D)(\phi)\,\delta_0(\phi)$ and $a_2\Phi = a_2(y,\nabla\phi)\,\delta_0'(\phi) + a_2(y,D)(\phi)\,\delta_0(\phi)$. By (iv), the function $a_2(y,\nabla\phi)$ vanishes in Γ, hence

$$A(y,D)(\Phi u) = (b(u) + au)\,\delta_0(\phi)\,,\ b = B(y,\nabla\phi,\nabla\cdot)\,,\ a = (a_2 + a_1)(\phi)\,.$$

According to (ii) and (iii), the product $F_x\delta_0(\phi)$ is well defined as a distribution in U with support in Γ. This yields

$$A(\Phi u)\,F_x = (b(u) + au)\,\delta_0(\phi)\,F_x,$$

where $\alpha \doteq a_1(\phi) + a_2(\phi)$. By (7.18) we obtain

$$u(x) = \int_U (b(u) + au)\,\delta_0(\phi)\,F_x.$$

We can eliminate the derivatives of u by partial integration taking into account that $\tau(\phi) = 0$:

$$u(x) = \int_U u(a - \operatorname{div} b - b)(F_x)\,\delta_0(\phi)\,\mathrm{d}y = \int_\Gamma u\,(a - \operatorname{div} b - b)\,F_x \frac{\mathrm{d}y}{\mathrm{d}\phi}. \quad (7.19)$$

The right-hand side does not obviously depend on derivatives of u and the second factor is a distribution in Γ. ▶

◀ Proof of Theorem 7.14. Take the function $\phi(x, r) = \rho(x) - r$. The compact $K \doteq \{(x, r); \phi(x, r) \geq 0\}$ is the set of all spheres $S \subset D$. Apply the formula (7.19) to the Darboux operator A, to the solution $u = Nf$ in the set K, to the function ϕ and to the fundamental solution F_n as in Proposition 7.19. We have

$$\begin{aligned} b &= -2r\left(\partial_r + \langle \nabla\rho, \nabla\rangle\right),\ a - \operatorname{div} b = 3 - n + r\Delta_y \rho, \\ -br_+^{n-2}(q + 0\imath)^{1-n} &= 2(n-2)\,r_+^{n-2}(q + 0\imath)^{1-n} \\ &\quad + 2(1-n)\left(r - \langle y - x, \nabla\rho\rangle\right) r_+^{n-1}(q + 0\imath)^{-n} \end{aligned}$$

and $\mathrm{d}y\mathrm{d}r/\mathrm{d}\phi = \mathrm{d}y$. ▶

◀ Proof of Theorem 7.15. Fix a point $x \in \operatorname{supp} f$ and draw a domain $U \subset \mathbf{E} \times \mathbb{R}_+$ that contains the point $(x, 0)$ as follows. Take an arbitrary vector $v \in \mathbf{E}\backslash\{0\}$ and consider the family of spheres $S(x + rv, r), r \geq 0$. The set $\mathbf{E}_x(v) \doteq \cup_{r\geq 0} S(x + rv, r)$ is equal to $\mathbf{E}$ if $|v| < 1$ and contains the cone $K_v \doteq \{x; (x-y, v) > \varepsilon|x - y|\}$ in the case $|v| \geq 1$ where $\varepsilon^2 = |v|^2 - 1$. If ε is small, the cone K_v is close to an open half-space whose boundary contains x. By assumption, the point x is in the interior $H(Y)$, which implies that the cone K_v meets Y for any v in the ball $B \doteq \left\{|v|^2 \leq 1 + \varepsilon^2\right\}$, if ε is sufficiently small. For small r the sphere $S(x + rv, r)$ does not meet Y, hence, for some r it is tangent to Y , i.e., $S(x + rv, r) \in \Sigma(Y)$. Let $\rho(v) > 0$ be the minimal r that possesses this property. The function $v \mapsto \rho(v)$ is continuous in B, moreover it is Lipschitz continuous. Consider the hypersurface $\Gamma(x) \doteq \{(y, r);\ y = x + \rho(v)v,\ r = \rho(v),\ v \in B\}$ in $\mathbf{E} \times \mathbb{R}_+$. It is contained in $\Sigma(Y)$, Lipschitz continuous and smooth except for a subset of zero measure area (centres of curvature of Y). Extend the function $\rho(v)$ to a positive continuous function in $\mathbf{E}$ that is smooth in $\mathbf{E}\backslash B$ and satisfies $\rho(v) = |v|^{-1}$ at infinity. Consider the domain

$$U(x) \doteq \text{closure}\left\{(y, r); 0 < r \leq \rho\left(\frac{y - x}{r}\right)\right\} \subset \mathbf{E} \times \mathbb{R}_+.$$

The intersection $U \cap \mathbf{E} \times \{0\}$ contains a 1-neighborhood of x and $\Gamma(x) \subset \partial U(x)$. The generalized function $F_x(y, r) = F_n(y - x, r)$ is the singular fundamental solution for the Darboux operator with source at x. This kernel fulfills the condition (ii), since the support of F_x is contained in the future cone $V + x$. This cone is the union of all rays emanating from x. The hypersurface $\Gamma(x)$ is transversal to the cone $V + x$ and we can see from explicit formulae of the next sections that the trace of F_x on $\Gamma(x)$ is well defined. This implies (iii). Thus all the conditions of Proposition 7.18 are fulfilled and we can apply the equation (7.19). ▶

Example 1. Let D be the unit ball in $\mathbf{E}^n$. By the above construction, we obtain the function $\rho(y) = 1 - |y|$ and cone $K = \{(y, r); r + |y| \leq 1\}$; the exceptional set C is empty. We have

$$\Phi_n(x, y, r) = -\frac{1}{2}\frac{\left(1 - |x|^2 - 2\langle y, x\rangle + 2|y||x|^2\right)}{(1 - |y|)|y|}.$$

By (7.16) for even n,

$$f(x) = -\frac{(n-1)!}{\mathrm{j}^n}\int_D \left[q^{-n}\right]\frac{1 - |x|^2 - 2\langle y, x\rangle + 2|y||x|^2}{1 - |y|}\frac{Mf(y, r)}{|y|}dy.$$

In the case n is odd the reconstruction looks similar to the function $(n-1)!\mathrm{j}^{-n}\left[q^{-n}\right]$ replaced by $2^{-1}\mathrm{j}^{1-n}\delta^{(n-1)}(q)$.

7.8 Fundamental solution for the adjoint operator

The adjoint Darboux operator is

$$A^* = r\frac{\partial^2}{\partial r^2} + (3 - n)\frac{\partial}{\partial r} - r\Delta_x.$$

We call a distribution F_n in $\mathbf{E} \times \mathbb{R}$ a *singular fundamental* solution for A^* if $A^*F_n = \delta_{0,0}$ where $\delta_{0,0} = \delta_{0,0}(x, r)$ is the Dirac distribution at the origin and $\operatorname{supp} F_n \subset \mathbf{E} \times \mathbb{R}_+$. We are going to find a singular fundamental solution for any n. We use the standard notation $r_+^\lambda = r^\lambda$, for $r > 0$, $\lambda \in \mathbb{C}$ and $r_+^\lambda = 0$ otherwise. Denote by $\mathrm{d}x$ the Euclidean volume form in $\mathbf{E}$.

Proposition 7.19 *Set* $q = \left(r^2 - |x|^2\right)/2$; *the function*

$$F_n = \left(-\frac{1}{2\pi}\right)^{(n-1)/2}\frac{1}{2(n-2)!!}r_+^{n-2}\delta^{(n-2)}(q) \tag{7.20}$$

for odd $n \geq 3$ *and the function*

$$F_n = \frac{(-1)^{n/2-1}(n-2)!}{2^{n-1}\pi^{n/2}\Gamma(n/2)}r_+^{n-2}\left[q^{1-n}\right] \tag{7.21}$$

for even $n \geq 2$ *is a singular fundamental solution for the Darboux operator. If* $n \geq 3$ *is odd,* F_n *is supported by the future cone.*

Lemma 7.20 *The function* $Q = r^{n-2}(q + 0\imath)^{1-n}$ *fulfils the equation* $A^*Q = 0$ *in* $E \times \mathbb{R}$.

◀ We have

$$A^*Q = r^{n-1}A\,(q+0\imath)^{1-n} = r^{n-1}\left(\partial_r^2 - \Delta_x + \frac{n-1}{r}\partial_r\right)(q+0\imath)^{1-n}.$$

The operator $\left(\partial_r^2 - \Delta_x\right)/2$ is dual to the quadratic form $q/2$ of $n+1$ variables x, r. Therefore by (1.10) we have $\left(\partial_r^2 - \Delta_x\right)(q+0\imath)^{1-n} = 2\,(n-1)^2\,(q+0\imath)^{-n}$. On the other hand, $(n-1)\,r^{-1}\,\partial_r\,(q+0\imath)^{1-n} = -2\,(n-1)^2\,(q+0\imath)^{-n}$ which implies that the right-hand side vanishes. ▶

◀ Now we prove the proposition. Take the function $Q_+ \doteq r_+^{n-2}\,(q+0\imath)^{1-n}$ and apply the operator

$$A^*Q_+ = A^*\left(r_+^0 Q\right) = 0 + 2r\delta_0\,(r)\,\partial_r Q + r\delta_0'\,(r)\,Q + (3-n)\,\delta_0\,(r)\,Q.$$

The second term vanishes, since the function $\partial_r Q$ is bounded. The third term is equal to $-2\delta_0\,(r)\,Q$ and $A^*Q = (1-n)\,\delta_0\,(r)\,Q$ where the product $\delta_0 Q$ can be evaluated as the limit $\delta\,(r-s)\,Q$ as $s \to +0$. We have for an arbitrary $\phi \mathrm{d}x \in K\,(\mathbf{E})$,

$$\begin{aligned}
\delta\,(r-s)\,Q\,(\phi) &= \int s^{n-2}\left(s^2 - |x|^2 + 0\imath\right)^{1-n}\phi\,(x)\,\mathrm{d}x \\
&= \int\left(1 - |y|^2 + 0\imath\right)^{1-n}\phi\,(sy)\,\mathrm{d}y \to c_n\phi\,(0)\,, \\
c_n &\doteq \int_{\mathbb{R}^n}\left(1 - |y|^2 + 0\imath\right)^{1-n}\mathrm{d}y
\end{aligned}$$

as $s \to 0$, where we change $y = x/s$. To calculate the last integral we use the spherical coordinates $y = t\omega$ and change again $t = -\imath v$. This reduces the integral to a Beta-function:

$$c_n = \frac{2\pi^{n/2}\,(-\imath)^n}{\Gamma\,(n/2)}\int_0^\infty \frac{v^{n-1}\mathrm{d}v}{(1+v^2)^{n-1}} = -\frac{\pi^{n/2}\,(-\imath)^n\,\Gamma\,(n/2-1)}{\Gamma\,(n-1)}.$$

Therefore $A^*Q_+ = c_n\delta_0\,(r)\,\delta_0\,(x) = c_n\delta_{0,0}\,(x,r)$, hence, $F_n \doteq \operatorname{Re} c_n^{-1}Q_+$ is a singular fundamental solution. The number c_n is imaginary for odd n and real for even n. Therefore $F_n = c_n^{-1}\operatorname{Re} Q_+$ for even n and $F_n = -\imath c_n^{-1}\operatorname{Im} Q_+$ for odd n. We have $\operatorname{Re} Q_+ = r_+^{n-2}\left[q^{1-n}\right]$, $\operatorname{Im} Q_+ = (-1)^{n-2}\,2\pi\,((n-2)!)^{-1}\,r_+^{n-2}\delta^{(n-2)}\,(q)$ by Section 1.4. Elementary calculations yield (7.20) and (7.21). ▶

Remark. In the case $n = 1$ the distribution $F_1\,(x,r) = (2r)_+^{-1}\,\theta\left(r^2 - x^2\right)\mathrm{d}x\mathrm{d}r$ is a singular fundamental solution for the Darboux operator.

Chapter 8

Algebraic Integral Transform

8.1 Problems

Let Y be a real vector space of dimension $n+1$ and m be a natural number; consider the space P_m of all non-trivial homogeneous polynomials of degree $m \geq 1$ with real coefficients in Y. Write a polynomial $a \in P_m$ as follows:

$$a(x) \equiv \sum_{|i|=m} a_i x^i, \text{ where } i = (i_0, \ldots, i_n) \in \mathbb{Z}_+^n.$$

The space is a real affine variety of dimension $N = \binom{m+n}{n}$. Any polynomial $a \in P_m$ defines the real algebraic cone $\{a(x) = 0\}$ in Y. Choose a coordinate system $x = (x_0, \ldots, x_n)$ in Y and consider the n-sphere $\mathbf{S}(Y) = \{\sum x_0^2 = 1\}$. Denote by $A \subset \mathbf{S}(Y)$ the intersection of this cone with the sphere; this is an algebraic variety in $\mathbf{S}(Y)$.

Take an even smooth differential form ω of degree $n-1$ on $\mathbf{S}(Y)$ and define the Funk transform of ω on P_m as follows:

$$F(\omega, a) = F(\omega, A) \doteq \int_A \omega, \quad a \in P_m. \tag{8.1}$$

Here we suppose that the form ω vanishes in a neighborhood of all singular points of A. Choose an orientation of Y; it generates the orientation of the sphere $\mathbf{S}(Y)$ and an orientation of A by means of the form $\mathrm{d}a$. We have $\mathrm{d}a(-x) = (-1)^m \mathrm{d}a(x)$. If $n+m$ is odd, the antipodal mapping $x \mapsto -x$ preserves the orientation of A and the projective variety $A_P \doteq A/\mathbb{Z}_2$ is oriented. Then the integral (8.1) is well defined for any $n-1$-form ω in the projective space $X \doteq \mathbb{P}(Y) = \mathbf{S}(Y)/\mathbb{Z}_2$. If $n+m$ is even, the projective variety A is not orientable. The integral (8.1) is defined for any even form ω in $X \doteq \mathbf{S}(Y)$. Note that in both cases $F(\omega, \lambda a) = \operatorname{sign} \lambda F(\omega, a)$ for any $\lambda \neq 0$. There are two problems of inversion for this transform:

Reconstruction problem: *to find the form* $\mathrm{d}\omega$ *from data of integrals* $F(\omega, A)$*for a sampling of polynomials* a *of degree* m. Note that in the case $\operatorname{supp}\omega \Subset \mathbb{R}^n$ the information contained in the data of $F(\omega, A)$ depends only on $\mathrm{d}\omega$, since the integrals vanish on exact forms $\omega = \mathrm{d}\nu$.

To avoid redundancy we need to assume that Γ is a pencil, that is an n-parametric family.

Torelli problem: *to recover the hypersurface* A *from knowledge of* $F(\omega, A)$ *for a sampling of forms* ω. These problems relate to the

Range problem: *to describe the space of functions* $a \mapsto F(\omega, a)$ that appear for various ω.

Denote by D_m the discriminant subset of P_m, that is the set of polynomials a such that A has, at least, one real singular point.

Proposition 8.1 *For any smooth form* ω *in* X *such that* $n + m + p(\omega)$ *is odd, the function* $F(\omega, A)$ *is smooth in* $P_m \backslash D_m$ *and has continuous extension to* P_m.

◀ The first statement is obvious. Take A_0, $A \in G$. Then $F(\omega, A) - F(\omega, A_0) = \int_B \mathrm{d}\omega$, where $B \subset X$ is the open set with boundary $\partial B = A - A_0$. The right side is bounded and depends continuously on B and therefore on A. This implies the second statement. ▶

8.2 Special cases

The above problems can be easily solved in the following simple cases:

HYPERPLANES. Take, first $m = 1$, $a(x) = \sum_0^n a^i x_i$. The Funk transform $F(\omega, a) = \int_{H(a)} \omega$ is well defined for hyperplanes $H(a) \doteq \{a(x) = 0\}$ in the projective space $\mathbb{P}^n$ for even n. The function F is odd: $F(\omega, -a) = -F(\omega, a)$ for any $n-1$-form ω. If n is odd, the hyperplane in $\mathbb{P}^n$ is not oriented and we define the Funk transform by the integral of an even form ω in the sphere $\mathbf{S}^n$ over the hyperplane $H(a) \subset \mathbf{S}^n$. The integral is again an odd function of a. Consider the inversion problem.

Proposition 8.2 *The form* $\mathrm{d}\alpha$ *can be uniquely reconstructed from data of* $F\alpha(H)$ *for all hyperplanes* $H \subset V$ *if the coefficients of the forms* α *and* $\mathrm{d}\alpha$ *are integrable in* Y.

◀ Let H be a hyperplane with an orientation and $V(H)$ be a half-space in Y bounded by H. We have by Stokes' thoerem

$$\int_{V(H)} \mathrm{d}\alpha = \int_H \alpha.$$

Choose a Euclidean structure in Y; let $\mathrm{d}Y$ be the volume element in this structure. Write $\mathrm{d}\alpha = \phi \mathrm{d}Y$ for a function ϕ and set $H = H(\omega, p)$. Taking p-derivative, yields

$$\frac{\partial}{\partial p} \int_{H(\omega,p)} \alpha = \int_{H(\omega,p)} \phi \mathrm{d}Y = R\phi(\omega, p).$$

We recover the function ϕ by means of inversion of the Radon transform. ▶

Now we discuss the range problem.

Theorem 8.3 *For an arbitrary smooth odd function g defined on the dual sphere $\mathbf{S}^*$, there exists a smooth $n-1$-form ω on $\mathbf{S}$ such that $F(\omega, a) = g$.*

◀ Any point $a \in \mathbf{S}^*$ defines the hemisphere $S(a) = \{\langle a, x\rangle \leq 0\}$ in $\mathbf{S}$. We want to solve the equation

$$\int_{S(a)} f(\alpha)\, \mathrm{d}S = g(a)$$

for an odd function f in $\mathbf{S}$. For this we apply the Funk-Hecke theorem to the function $\theta(t) = t_+^0$:

$$\int_{\mathbf{S}} \theta(a\alpha)\, Y_l(\alpha)\, \mathrm{d}\alpha = c(n,l)\, Y_l(a)\,, \ a \in \mathbf{S}^*,$$

$$c(n,l) = \frac{2\pi^{n/2}}{\Gamma(n/2)} \int_0^1 C_l^{(n-1)/2}(t)\left(1-t^2\right)^{(n-2)/2} \mathrm{d}t,$$

where Y_l are spherical harmonics. By [93] we find

$$c(n, 2k+1) = MG_{2k+1}^{\lambda}(1) = \frac{\pi^{1/2}\Gamma(\lambda+1/2)}{2\Gamma(k+\lambda+3/2)\Gamma(1/2-k)},$$

where G^λ are Gegenbauer polynomials of degree $\lambda = (n-1)/2$ and M is the Mellin transform. We have $c(n, 2k+1) \sim C_n k^{-(n+1)/2}$ as $k \to \infty$. Write

$$g(a) = \sum g_k Y_{2k+1}(a)\,.$$

The odd function

$$f(\alpha) = \sum \frac{g_k}{c(n, 2k+1)} Y_{2k+1}(\alpha)$$

solves the above equation. The coefficient $g_k / c(n, 2k+1)$ decreases fast as $k \to \infty$ since g_k does so. Therefore $f \in C^\infty(\mathbf{S}^n)$ and

$$\int_{\mathbb{S}^n} f \mathrm{d}S = g(a) + g(-a) = 0.$$

Therefore there exists an even smooth $n-1$-form ω in $\mathbf{S}^n$ such that $\mathrm{d}\omega = f\mathrm{d}S$. We have

$$\int_{H(a)} \omega = \int_{\partial S(a)} \omega = \int_{S(a)} f \mathrm{d}S = g(a)\,.$$ ▶

This theorem solves the range problem for $m = 1$. The case $m > 1$ is open.

Linear pencil. Another simple case is the pencil $\Sigma(F) \subset P_m$ of hypersurfaces which contain a fixed set F of $p \doteq N-1-n$ points $x_1, \ldots, x_p$. This is a linear pencil in $P_{n,m}$ of dimension n and therefore is equal to the envelope of $n+1$ hypersurfaces, say $A_0, \ldots, A_n$. This means that any $A \in \Sigma$ is given by the equation $a = 0$, where $a \doteq c_0 a_0 + \cdots + c_n a_n$, and $a_i = 0$ is an equation of A_i. Now we change homogeneous coordinates to affine coordinates, for example $y_i = a_i(x)/a_0(x)$, $i = 1, \ldots, n$, in the set $U \doteq \mathbb{P}^n \backslash A_0 \cup D$, where D is the zero set of the Jacobian. Then any element of Σ may be given by the equation

$$1 + \sum \xi_i y_i = 0, \quad (\xi_1, \ldots, \xi_n) \in \mathbb{R}^n.$$

Hence we can reconstruct any function $f \in C_0(U)$ by the methods of Chapter 2. If a function f does not vanish on D, we can find its values by means of integral asymptotics.

As an example we consider the integrals of a form ω over a pencil of conics. Take a family of quadratic hypersurfaces $q(x) = 0$ in V. Fix N points $p_1, \ldots, p_N \in V$ in general position, where $N = n(n+1)/2$ and take the family Γ_N of quadrics that contain these points: $q(p_j) = 0, j = 1, \ldots, N$. The family of quadratic functions q satisfying these equations is a vector space Q of dimension $n + 1$, hence the family Γ_N has n parameters.

We show that the data $F\omega|\Gamma_N$ is sufficient for reconstruction, at least, locally. Choose a basis $q_0, \ldots, q_n$ in Q; it defines the algebraic mapping $q : V \to \mathbb{R}^{n+1}$, $\xi_j = q_j(x)$, $j = 0, \ldots, n$. Consider the projective space $\mathbb{P}^n = \mathbb{P}(\mathbb{R}^{n+1})$ and choose an affine chart, say, the chart $\mathbf{W} \cong \mathbb{R}^n$ with coordinates $y_1 = \xi_1/\xi_0, \ldots, y_n = \xi_n/\xi_0$. The mapping $q : V \backslash \{q_0 = 0\} \to \mathbf{W}$ is well defined; choose an open subset $V' \subset V \backslash (q_0 = 0)$ such that the mapping $V' \to \mathbf{W}$ is an embedding. This means that the functions $y_j = q_j(x)/q_0(x), j = 1, \ldots, n$ are coordinates in V'. Any linear equation $a_0 + a_1 y_1 + \cdots + a_n y_n = 0$ is equivalent to $a_0 q_0 + \cdots + a_n q_n = 0$, i.e., any hyperplane H belongs to the family Γ_N. For an arbitrary form ω of degree $n - 1$ such that $\operatorname{supp} \omega \subset V'$, all the hyperplane integrals are known. According to Proposition 8.2 the form $d\omega$ can be reconstructed from this data.

WERTGEIM'S PENCIL. Following [97], we consider the case where $n = m = 2$, and the pencil Σ of conic curves which are tangent to a given conic curve B in two points and pass through a fixed point x_0. Then we take the double covering $\sigma : \mathbb{D} \to \mathbb{P}^2$, defined by the equation $z = \sqrt{b(x)}$, where $b(x) = 0$ is an equation of B. This covering is in fact a conic $\mathbb{D}$ in $\mathbb{P}^2$ and any curve $A \in \Sigma$ is the projection to $\mathbb{P}^2$ of the intersection $\mathbb{D} \cap H$, where H is a plane containing the point $z_0 = (x_0, \sqrt{b(x_0)})$. We identify any point $z \in \mathbb{D}$ with the line through z_0 and z, thus we get the canonical birational isomorphism $\mathbb{D} \simeq \mathbb{P}^2$ such that plane sections of $\mathbb{D}$ correspond to projective lines. Taking an unknown form ω on $\mathbb{P}^2$, we lift it to a form $\sigma^*(\omega)$ on $\mathbb{D}$ and then to $\mathbb{P}^2$. Thus the data $F(\omega, A)$ coincide with the Radon data of $\sigma^*(\omega)$ on the projective plane. Then we apply the method of the first case.

8.3 Multiplicative differential equations

We now return to the general situation of Section 8.1.

Theorem 8.4 *The function $F = F(\omega, \cdot)$ satisfies the following system of differential equations in $P_m \backslash D_m$:*

$$\sum a_i \frac{\partial}{\partial a_i} F = 0, \tag{8.2}$$

$$\left(\frac{\partial}{\partial a_i}\frac{\partial}{\partial a_j} - \frac{\partial}{\partial a_k}\frac{\partial}{\partial a_l}\right) F = 0, \text{ if } i + j = k + l. \tag{8.3}$$

Characteristic variety of a differential system. Consider a system of linear partial differential equations

$$a_j(x, D)\, u = 0, \; j = 1, \ldots, s \tag{8.4}$$

with smooth coefficients in an open set $U \subset V$ and one unknown function u. Fix an integer k and consider the set of differential operators of order $\leq k$ of the form

$$a(x, D) = \sum b_j(x, D)\, a_j(x, D),$$

where b_j are arbitrary differential operators with smooth coefficients. Take the principal part A_k of order k of the operator a and consider it as a function $A_k(x, \xi)$ on the complex cotangent bundle $T^*_{\mathbb{C}}(U)$. It is a smooth function of x with a homogeneous polynomial of degree k in the variables $\xi = (\xi_1, \ldots, \xi_n)$, in the fibre T^*_x of the cotangent bundle, $T^*_x \simeq \mathbb{C}^n$, $n = \dim V$. Let N be the set of common zeros of functions $A_k(x, \xi)$, $k = 0, 1, 2, \ldots$. It is called the *characteristic variety* of the system (8.4). This is an analytic set in $T^*(U)$; the fibre $N_x \doteq \pi^{-1}(x)$ of the mapping $\pi : N \to U$ is a complex algebraic cone in $\mathbb{C}^n$.

The equation (8.2) expresses the fact that F is a homogeneous function of degree zero. Describe the characteristic variety of this system. Let R be the real line dual to $\mathbb{R}$; C be the dual to $\mathbb{C}$. Consider the Veronese mapping Ver_m: $\mathrm{C}^{n+1} \to \mathrm{C}^N$ given by the equations $\xi_i = \eta^i$, $i = (i_0, \ldots, i_n)$, $\eta \in \mathbb{C}^{n+1}$, or symbolically $\xi = \eta^m$.

Proposition 8.5 *The characteristic variety of the system* (8.3) *in P_m is equal to* $P_m \times \mathrm{Ver}_m\left(\mathrm{C}^{n+1}\right)$.

◀ The characteristic variety is defined by the system of quadratic equations $\xi_i \xi_j - \xi_k \xi_l = 0$, for any i, j, k, l such that $i + j = k + l$. This system implies that $\xi_i = \eta^i$ for some $\eta \in \mathrm{C}^{n+1}$. ▶

The characteristic variety V of the system (8.2–8.3) is an algebraic manifold in the complex cotangent bundle $T^*_{\mathbb{C}}(P_m) = P_m \times \mathrm{C}^N$ of the real algebraic manifold P_m. The characteristic variety of (8.2) equals the hyperplane $H \doteq \{a \in P_m,\ \xi \in \mathrm{C}^N;\ \langle a, \xi\rangle \equiv \sum a_i \xi_i = 0\}$. The characteristic variety $J \subset T^*_{\mathbb{C}}(P_m)$ of the system (8.2–8.3) is equal to the intersection of $\left(P_m \times \mathrm{Ver}_m\left(\mathrm{C}^{n+1}\right)\right) \cap H$.

Corollary 8.6 *For an arbitrary point $a \in P_m$, the fibre J_a of the characteristic variety J is equal to the variety*

$$\mathrm{Ver}_m\left(\mathrm{C}^{n+1}\right) \cap H_a = \left\{\xi;\ \xi_i = z^i, \sum a_i z^i = 0\right\},$$

which is isomorphic to the complex variety $A_{\mathbb{C}}$ given in the parametric form $\xi_i = z^i$, $|i| = m$.

It follows that the system (8.2–8.3) is maximal, that is any linear differential equation that is fulfilled by all the functions $F(\omega, \cdot)$ is a corollary of this system. Moreover, this provides a method for solving the Torelli problem: if we know enough functions of the form $F(\omega, \cdot)$ in a neighborhood of a point a to be identified, we can recover the system of equations they satisfy. Taking the characteristic variety of this system, we reconstruct the characteristic manifold and the fibre J_a which gives the variety $A_{\mathbb{C}}$.

Example 1. Let $m = 2$; the variety $\mathrm{Ver}_m\left(\mathrm{C}^{n+1}\right)$ is equal to the intersection of $n(n+1)/2$ quadrics. This implies that the function F satisfies a system of second degree differential equations.

Example 2. Any solution y of the algebraic equation

$$a_m y^m + a_{m-1} y^{m-1} + \cdots + a_0 = 0, \quad m > 1,$$

is an analytic function of the coefficients $a = \{a_m, \dots, a_0\}$ in the set $\mathbb{R}^{m+1} \backslash D$ where D is the discriminant set of the polynomial $a(t) \doteq a_m t^m + \cdots + a_0$. Then y^k satisfies the equations (8.2–8.3) for $k = 1, 2, \dots$ and also the nonlinear equations are fulfilled

$$\frac{\partial y}{\partial a_i}\frac{\partial y}{\partial a_j} = \frac{\partial y}{\partial a_k}\frac{\partial y}{\partial a_l}, \quad i + j = k + l,$$

which coincide with the Eikonal equations for (8.3).

Proposition 8.7 *For any point $(a, \xi) \in \mathrm{Re}\, J$ there exists a sequence of smooth forms $\omega_t, t = 1, 2, \dots$ in X such that $F(\omega_t) \to v$ in $D'(P_m \backslash D_m)$ where v is a distribution-solution of (8.2–8.3), such that $(a, \xi) \in WF(v)$.*

◀ We have $\xi = \mathrm{Ver}_m(\eta)$ for some $\eta \in \mathbb{R}^{n+1} \backslash 0$, $a(\eta) = 0$, and take $\omega_t(x) = |x|^m \delta_t / \mathrm{d}_x a$, where δ_t, $t \to \infty$ is a sequence of smooth distributions on X, which tends to the Dirac distribution at the point $x = \eta$ and $\mathrm{supp}\, \delta_t \to \{\eta\}$. The function sequence $F(\omega_t)$ converges to a generalized function u on P_m which is supported by the hypersurface $S = \{b \in P_m;\ b(\eta) = 0\}$. Hence $WF(u) \supset N^*(S) = \{(b, \xi); \langle b, \xi \rangle = 0, \xi = \eta^m\}$. This implies that the set $WF(u)$ contains the covector (a, ξ). ▶

This assertion gives a method to solve the Torelli problem: if we know the function $F(\omega, A)$ for an unknown variety A but for sufficiently many forms ω, we can recover the variety $\mathrm{Re}\, J_a$ which is isomorphic to the cone over A.

Problem: Is (8.2–8.3) the complete system of conditions for the range of the Funk transform?

The positive answer would mean that for any smooth function F defined in a neighborhood of a point $a_0 \in P_m \backslash D_m$ that satisfies this system, there exists a smooth $n-1$-form ω in a neighborhood of $\mathbb{P}(A_0)$ such that $F = F(\omega, a)$ for all a close to a_0. Note that the equations of (8.3) have constant coefficients and the exponential representation works for solutions of this system, see [66]. On the other hand, the equation (8.2) looks simple, but it makes the system dependent on a. The positive answer to the above problem seems plausible, at least, for non-singular varieties A. If A_0 has a singular point y and ω does not vanish at y, the function $F(\omega, A)$ is (at least, generically) singular at A_0. The simplest class consists of singular points of multiplicity 1 (Morse singularities). For this class, singularities of the integral F are described in Section 3.9.

8.4 Funk transform of Leray forms

Let Ω be a smooth n-form in an oriented n-manifold X. The quotient $\upsilon = \Omega / \mathrm{d}_x a$ is a form such that $\mathrm{d}_x a \wedge \upsilon = \Omega$. This quotient (called a Leray form) is well defined up to a term $\mathrm{d}a \wedge \Psi$ in a neighborhood of a non-singular hypersurface $A = \{a(x) = 0\}$. Choose the orientation of A by means of $\mathrm{d}a$. The integral

$$F(\upsilon, A) = \int_A \upsilon$$

is uniquely defined if $\operatorname{supp} \upsilon \cap A$ is a compact set. The function $F(\upsilon, A)$ is well defined for $A \in P_m \backslash D_m$. It is a smooth (analytic) function of A if so is the form Ω.

Proposition 8.8 *For any smooth $n-1$-form ω we have*

$$\frac{\partial}{\partial a_0} \int_A \omega = \int_A \frac{\mathrm{d}\omega}{\mathrm{d}_x a} \tag{8.5}$$

for any $a \in P_m \backslash D_m$.

◄ Let $a(x) = \sum a_j x^j$; set $A(t) = \left\{\sum a_j(t) x^j = 0\right\}$ where $a_j(t) = a_j$, except for $j = 0$, and $a_0(t) = a_0 + t$ for small t. Then $F(\omega, A(t)) - F(\omega, A) = \int_{B(t)} \mathrm{d}\omega$, where $B(t) \doteq \{0 \geq a(x) \geq -t\} \subset X$ for $t > 0$ and $B(t) = \{-t \geq a(x) \geq 0\}$ for $t < 0$. We have $\partial B(t) = A(t) - A$. The right side is bounded and depends continuously on t. By Fubini's theorem,

$$\int_{B(t)} \mathrm{d}\omega = \int_{-t}^{0} \mathrm{d}s \int_{A(s)} \frac{\mathrm{d}\omega}{\mathrm{d}_x a}.$$

Taking the derivative yields

$$\frac{\mathrm{d}}{\mathrm{d}t} F(\omega, a(t)) \bigg|_{t=0} = \frac{\mathrm{d}}{\mathrm{d}t} \int_{B(t)} \mathrm{d}\omega \bigg|_{t=0} = \int_A \frac{\mathrm{d}\omega}{\mathrm{d}_x a}. \qquad ►$$

Corollary 8.9 *The integral of the Leray form fulfils*

$$\sum a_i \frac{\partial}{\partial a_i} F(v,a) + F(v,a) = 0, \tag{8.6}$$

$$\left(\frac{\partial}{\partial a_i}\frac{\partial}{\partial a_j} - \frac{\partial}{\partial a_k}\frac{\partial}{\partial a_l}\right) F(v,a) = 0, \text{ if } i+j = k+l. \tag{8.7}$$

◀ We can write $\Omega = \mathrm{d}\omega$ in a neighborhood of $\operatorname{supp}\Omega \cap A$ for a smooth form ω; apply (8.5) and (8.2–8.3) to ω. By commuting with $\partial/\partial a_0$ we get (8.6–8.7). ▶

8.5 Differential equations for hypersurface integrals

Theorem 8.10 *Let X be an n-manifold, f be a smooth function in $X \times \Xi$ where $\Xi \subset \mathbb{R}^m$ is open, such that $\mathrm{d}_x f \neq 0$. Then for any smooth $n-1$-form ω in X with compact support and any point $\xi \in \Xi$ the hypersurface integral*

$$I(\omega, \xi) = \int_{f(\cdot,\xi)=0} \omega$$

oriented by the form $\mathrm{d}_x f$, satisfies the second-order equation

$$P(\xi, D_\xi) I \doteq \sum p_{jk}(\xi) \frac{\partial^2 I}{\partial \xi_j \partial \xi_k} = 0 \tag{8.8}$$

with some smooth coefficients p_{jk} in Ξ provided

$$P(\xi, \nabla_\xi f) \doteq \sum p_{jk}(\xi)\, frac\partial f(x,\xi) \partial \xi_j \frac{\partial f(x,\xi)}{\partial \xi_k} = 0 \tag{8.9}$$

and

$$P(\xi, D_\xi) f \; - \frac{\mathrm{d}_x P(\xi, \nabla_\xi f)}{\mathrm{d}_x f} = 0 \tag{8.10}$$

in $\mathsf{F} \doteq \{(x,\xi); f(x,\xi) = 0\}$.

Corollary 8.11 *If $P(\xi, D_\xi f) = 0$ in F and $P(\xi, \nabla_\xi) f = 0$ in $X \times \Xi$, then $P(\xi, D_\xi) I = 0$.*

Remark. The sufficient conditions of the corollary are not invariant under replacing f by $g = hf$, $h \neq 0$, whereas the conditions of Theorem 8.10 are.

◀ Proof of Theorem 8.4. The function $f(x,\xi) = a(x)$ is linear in $\xi_i = a_i$, $|i| = m$, which yields $P(\xi, D_\xi) f = 0$ for an arbitrary second-order operator P. The equation $\partial f/\partial \xi_i = x^i$ implies the system of equations in $X \times \Xi$:

$$P_{ijkl}(\xi, \nabla_\xi f) \doteq \frac{\partial f}{\partial \xi_i}\frac{\partial f}{\partial \xi_j} - \frac{\partial f}{\partial \xi_k}\frac{\partial f}{\partial \xi_l} = 0, \; i+j = k+l,$$

which have the form (8.9). The equations $\mathrm{d}_x P_{ijkl}/\mathrm{d}_x f = 0$ follow. Therefore (8.8) gives (8.3). ▶

Lemma 8.12 *Let ω be an $n-1$-form in $X \times \Xi$ such that $\mathrm{d}\omega = \sum \mathrm{d}\xi_i \wedge \omega_i$ in F for some smooth forms $\omega_i, |i| = m$ in $X \times \Xi$. Then*

$$\frac{\partial}{\partial \xi_i} I(\omega, \xi) = I(\omega_i, \xi). \tag{8.11}$$

◀ Proof of the lemma. Choose a point η close to ξ and an oriented curve γ that joins ξ and η. The n-chain $\mathsf{F}(\xi, \eta) = \{\mathsf{F}(\xi(t)), \xi(t) \in \gamma, 0 \le t \le 1\} \subset \mathsf{F}$ satisfies $\partial \mathsf{F}(\xi, \eta) = (\mathsf{F}(\eta), \eta) - (\mathsf{F}(\xi), \xi)$ and

$$\begin{aligned}\int_{(\mathsf{F}(\eta),\eta)} \omega - \int_{(\mathsf{F}(\xi),\xi)} \omega = \int_{\mathsf{F}(\xi,\eta)} \mathrm{d}\omega = \sum \int_{F(\xi,\eta)} \mathrm{d}\xi_i \wedge \omega_i = \sum \int_0^1 \mathrm{d}\xi_i(t) \int_{\mathsf{F}(\xi(t))} \omega_i \\ = \sum (\eta_i - \xi_i) \int_{\mathsf{F}(\xi)} \omega_i + o(|\xi - \eta|),\end{aligned}$$

which implies (8.11) as $\eta \to \xi$. ▶

◀ Proof of Theorem 8.10. By applying a partition of unity to ω, we reduce the proof to the case when ω is supported by a compact set $K \subset X$ covered by a local coordinate system $x_1, x_2, \ldots, x_n$ such that $\partial f(x, \xi) / \partial x_1 \neq 0$ for ξ in an open set $V \subset \Xi$. Write $\mathrm{d}\omega = \phi \mathrm{d}x_1 \wedge \mathrm{d}x'$; $\mathrm{d}x' \doteq \mathrm{d}x_2 \wedge \cdots \wedge \mathrm{d}x_n$, where ϕ is a smooth function in K, and set $\upsilon \doteq g^{-1} \phi \mathrm{d}x'$, where $g \doteq \partial f / \partial x_1$. Then

$$\mathrm{d}\omega = \mathrm{d}_x f \wedge \upsilon = -\mathrm{d}_\xi f \wedge \upsilon = -\sum \mathrm{d}\xi_j \wedge \frac{\partial f}{\partial \xi_j} \upsilon \tag{8.12}$$

in F since $0 = \mathrm{d}f = \mathrm{d}_x f + \mathrm{d}_\xi f$. This yields by the lemma

$$\frac{\partial I(\omega, \cdot)}{\partial \xi_j} = -I\left(\frac{\partial f}{\partial \xi_j} \upsilon, \cdot\right), \ j = 1, 2, \ldots, m. \tag{8.13}$$

By (8.12) for any j

$$\mathrm{d}\left(\frac{\partial f}{\partial \xi_j} \upsilon\right) = g^{-1} \frac{\partial^2 f}{\partial x_1 \partial \xi_j} \mathrm{d}_x f \wedge \upsilon + \sum_k \frac{\partial^2 f}{\partial \xi_k \partial \xi_j} \mathrm{d}\xi_k \wedge \upsilon + \frac{\partial f}{\partial \xi_j} \mathrm{d}\upsilon, \tag{8.14}$$

where

$$\begin{aligned}\mathrm{d}\upsilon &= \frac{\partial \left(g^{-1} \phi\right)}{\partial x_1} \mathrm{d}x_1 \wedge \mathrm{d}x' + \mathrm{d}_\xi g^{-1} \wedge \phi \mathrm{d}x' \\ &= \mathrm{d}_x f \wedge g^{-1} \frac{\partial \left(g^{-1} \phi\right)}{\partial x_1} \mathrm{d}x' - g^{-2} \sum_k \frac{\partial^2 f}{\partial \xi_k \partial x_1} \mathrm{d}\xi_k \wedge \phi \mathrm{d}x'.\end{aligned}$$

We have $\mathrm{d}_x f = -\mathrm{d}_\xi f$ in F, which yields

$$\mathrm{d}\upsilon = -\sum \frac{\partial f}{\partial \xi_k} \mathrm{d}\xi_k \wedge \psi - g^{-1} \sum \frac{\partial^2 f}{\partial \xi_k \partial x_1} \mathrm{d}\xi_k \wedge \upsilon,$$

where $\psi \doteq g^{-1}\partial\left(g^{-1}\phi\right)/\partial x_1 \mathrm{d}x' = \mathrm{d}_x\upsilon/\mathrm{d}_x f$. Substituting in (8.14) yields

$$\begin{aligned}
\mathrm{d}\left(\frac{\partial f}{\partial\xi_j}\upsilon\right) &= -g^{-1}\left[\frac{\partial^2 f}{\partial x_1\partial\xi_j}\sum_k\frac{\partial f}{\partial\xi_k}\mathrm{d}\xi_k + \frac{\partial f}{\partial\xi_j}\sum\frac{\partial^2 f}{\partial x_1\partial\xi_k}\mathrm{d}\xi_k\right]\wedge\upsilon \\
&\quad+\sum_k\frac{\partial^2 f}{\partial\xi_k\partial\xi_j}\mathrm{d}\xi_k\wedge\upsilon - \frac{\partial f}{\partial\xi_j}\frac{\partial f}{\partial\xi_j}\mathrm{d}\xi_k\wedge\psi \\
&= \sum_k \mathrm{d}\xi_k\wedge\left(-g^{-1}\frac{\partial}{\partial x_1}\left(\frac{\partial f}{\partial\xi_j}\frac{\partial f}{\partial\xi_k}\right)\upsilon + \frac{\partial^2 f}{\partial\xi_k\partial\xi_j}\upsilon - \frac{\partial f}{\partial\xi_j}\frac{\partial f}{\partial\xi_k}\psi\right) \\
&= \sum_k \mathrm{d}\xi_k\wedge\left(-\frac{\mathrm{d}_x\left(\frac{\partial f}{\partial\xi_j}\frac{\partial f}{\partial\xi_k}\right)}{\mathrm{d}_x f}\upsilon + \frac{\partial^2 f}{\partial\xi_k\partial\xi_j}\upsilon - \frac{\partial f}{\partial\xi_j}\frac{\partial f}{\partial\xi_k}\psi\right),
\end{aligned} \tag{8.15}$$

where we take into account the equation

$$g^{-1}\frac{\partial}{\partial x_1}\left(\frac{\partial f}{\partial\xi_j}\frac{\partial f}{\partial\xi_k}\right) = \frac{\mathrm{d}_x\left(\frac{\partial f}{\partial\xi_j}\frac{\partial f}{\partial\xi_k}\right)}{\mathrm{d}_x f}.$$

Therefore by (8.13) and by (8.15)

$$\frac{\partial^2 I(\omega,\cdot)}{\partial\xi_i\partial\xi_k} = -\frac{\partial}{\partial\xi_k}I\left(\frac{\partial f}{\partial\xi_j}\upsilon,\cdot\right) = I\left(\left[\frac{\mathrm{d}_x\left(\frac{\partial f}{\partial\xi_j}\frac{\partial f}{\partial\xi_k}\right)}{\mathrm{d}_x f} - \frac{\partial^2 f}{\partial\xi_k\partial\xi_j}\right]\upsilon + \frac{\partial f}{\partial\xi_j}\frac{\partial f}{\partial\xi_k}\psi,\cdot\right),$$

where $\upsilon = \mathrm{d}\omega/\mathrm{d}_x f, \psi = \mathrm{d}_x\upsilon/\mathrm{d}_x f$. Multiplying by p_{jk} and taking the sum yields

$$\sum p_{jk}\frac{\partial^2 I}{\partial\xi_i\partial\xi_k} = \sum P(\xi,\nabla_\xi f)\psi + \left[\frac{\mathrm{d}_x P(\xi,\nabla_\xi f)}{\mathrm{d}_x f} - P(\xi,D)f\right]\upsilon,$$

we complete the proof. ▶

Remark. The above construction can be generalized to the case of several smooth functions $f_1,\dots,f_k$ defined in $X\times\Xi$. For an $n-k$-form ω with compact support in X, the integral

$$I(\omega,\xi) = \int_{F(\xi)}\omega$$

can be defined over the manifold $F(\xi) = \{f_1(\cdot,\xi) = \dots = f_k(\cdot,\xi) = 0\}$ provided the forms $\mathrm{d}f_1,\dots,\mathrm{d}f_k$ are independent at each point of $F\cap\operatorname{supp}\omega$. The function I satisfies a more complicated system of equations, like that in Theorem 8.10.

8.6 Howard's equations

Consider the family of quadratic hypersurfaces $Q = Q(a,y,t)$ in X defined as

$$Q(a,y,t) \doteq \sum a_{ij}(x_i-y_i)(x_j-y_j) - t = 0.$$

Take a continuous function F in E with compact support and integrate it over Q:

$$g(a,y,t) = \int_{Q=0} f\mathrm{d}\omega$$

against the "affine invariant" element $\mathrm{d}\omega$ that is equal to the Euclidean surface area in any coordinate system x' such that Q is the unit sphere, see [48]. The function g satisfies a system of differential equations with respect to the variables $\xi = (a,y,t) \in \mathbb{R}^N$ where $N = (n+1)(n+2)/2$. One equation follows from the homogeneity of the quadratic function Q:

$$\left(\sum a_{ij}\frac{\partial}{\partial a_{ij}} + t\frac{\partial}{\partial t}\right) g = 0. \tag{8.16}$$

Now take the coordinates c_{ij} instead of a_{ij}, where $\{c_{ij}\} = \{a_{ij}\}^{-1}$. Howard's identity for elliptic Q reads

$$\frac{\partial}{\partial c_{ij}} g(a,y,d) = \frac{d^{1-n/2}}{2}\int_0^b t^{n/2-1}\frac{\partial}{\partial y_i}\frac{\partial}{\partial y_j} g(a,y,t)\,\mathrm{d}t,\ i,j = 1,\dots,n.$$

It is equivalent to

$$2\frac{\partial^2}{\partial t\partial c_{ij}} h(a,y,t) = \frac{\partial^2}{\partial y_i\partial y_j} h(a,y,t), \tag{8.17}$$

where $h(a,y,t) = t^{n/2-1} g(a,y,t)$.

Corollary 8.13 *The equations* (8.17) *hold for an arbitrary quadratic* Q.

◀ First, we check that the function Q fulfils

$$4\frac{\partial Q}{\partial c_{ij}}\frac{\partial Q}{\partial t} = \frac{\partial Q}{\partial y_i}\frac{\partial Q}{\partial y_j}, \tag{8.18}$$

which is of the form (8.9). This follows from the equations

$$\frac{\partial Q}{\partial c_{ij}} = -\sum_{kl} a_{ik}(x_k - y_k)\,a_{jl}(x_l - y_l),\ \frac{\partial Q}{\partial t} = -1,\ \frac{\partial Q}{\partial y_i} = 2\sum_k a_{ik}(x_k - y_k).$$

This yields that the function $R \doteq c^{-1/2}Q,\ c = \det\{c_{ij}\}$ also satisfies (8.18) in the hypersurface $R = 0$. We have

$$\frac{\partial^2 R}{\partial t\partial c_{ij}} = \frac{\partial c^{-1/2}}{\partial c_{ij}}\frac{\partial Q}{\partial t} = c^{-3/2}\frac{b_{ij}}{2},\ \frac{\partial^2 R}{\partial y_i\partial y_j} = c^{-1/2}\frac{\partial^2 Q}{\partial y_i\partial y_j} = c^{-1/2}2a_{ij},$$

where $b_{ij} = ca_{ij}$ is the complement minor to the element c_{ij} in the matrix $\{c_{kl}\}$. This implies the equation

$$4\frac{\partial^2 R}{\partial t\partial c_{ij}} = \frac{\partial^2 R}{\partial y_i\partial y_j}.$$

This together with (8.18) yields (8.17) by Theorem 8.10. ▶

We want to exclude the derivatives with respect to c_{ij} for the family of quadrics Q with a fixed homogeneous part $Q(a,y,t) \doteq a^{ij}(x_i - y_i)(x_j - y_j) - t$, $a^{ij} = a^{ji} = \text{const}$. Instead of (8.16), we have

$$\sum c_{ij}\frac{\partial h}{\partial c_{ij}} = -\sum a_{ij}\frac{\partial h}{\partial a_{ij}} = t\frac{\partial h}{\partial t} + \left(\frac{n}{2} - 1\right)h.$$

Multiply (8.17) by c_{ij} and take the sum:

$$4\frac{\partial}{\partial t}\left(t\frac{\partial}{\partial t} + \left(\frac{n}{2} - 1\right)\right)h = \sum c_{ij}\frac{\partial^2}{\partial y_i \partial y_j}h.$$

Set $t = r^2$ and get

$$\frac{\partial^2 h}{\partial r^2} + \frac{n-1}{r}\frac{\partial h}{\partial r} = \sum c_{ij}\frac{\partial^2 h}{\partial y_i \partial y_j},$$

which is the Darboux equation for conics of arbitrary shape. In the case $a_{ij} = c_{ij} = \delta_i^j$, the function h equals the spherical integral of f.

8.7 Range of differential operators

Let V^n be a vector space; the variety $A_{n-1}(V)$ of affine hyperplanes $H \subset V$ has a structure of affine algebraic manifold. Moreover, the variety $A_{n-1}(V)$ is the affine part of the closed algebraic manifold $\mathbb{P}^n$ that is dual to the projective completion of V. Suppose now that V has Euclidean structure and W is an algebraic submanifold of $A_{n-1}(V)$. If f is a function in V with compact support and we know the integrals $Mf(H)$ for only $H \in W$, which information on f can be reconstructed?

Example 3. Let $n = 1$ and we know only $\int f\mathrm{d}x$. Then we can reconstruct the function f modulo the subspace of functions g whose integral over V vanishes, i.e., modulo the space of functions $g = h'$ where h is again a function with compact support.

Example 4. Let $n = 2$ and suppose that we know all the integrals $Mf(H)$ for lines H parallel to x-axes or to y-axes. Then we can recover f modulo the subspace of functions of the form h_{xy} where h has compact support.

Let now $\mathbf{V}$ be a vector space of dimension n and $q = q(\xi)$ be a homogeneous polynomial in the dual space V^*. The integral $M[q(D)f](H)$ vanishes for any hyperplane $H = H(\omega, p)$ such that $q(\omega) = 0$, since integrating by parts yields for any $\lambda \in \mathbb{R}$,

$$\int \exp(\imath\lambda p)\,\mathrm{d}p \int_{H(\omega,p)} q(D) f\mathrm{d}S = \int \exp(\imath\lambda(\omega, x))\, q(D) f\mathrm{d}x$$
$$= q(-\imath\lambda\omega)\int \exp(\imath\lambda(\omega,x)) f\mathrm{d}x = 0.$$

This implies that $Mq(D) f(H) = 0$. The inverse is also true. Let $W \subset V^*$ be a real conic algebraic variety and $I(W)$ be the ideal of polynomials in V^* that vanishes on W. Let $q_1, \ldots, q_r$ be a set of homogeneous generators of the ideal $I(W)$.

Theorem 8.14 *If all the integrals $Mf(H(\omega, p))$, $\omega \in W$, $p \in \mathbb{R}$ are known for a smooth function f with compact support, then this function can be recovered modulo the subspace of functions of the form $q_1(D) g_1 + \cdots + q_r(D) g_r$ where $g_1, \ldots, g_r$ are smooth functions with compact support.*

Example 5. Let $\dim V = 4$ and W be given by $q(\xi) \doteq \xi_0^2 - \xi_1^2 - \xi_2^2 - \xi_3^2 = 0$. The ideal $I(W)$ is generated by q. Therefore we can recover the function f from data of integrals $Mf(H(\omega, p)), q(\omega) = 0$ modulo functions of the form $\Box g$. In particular, for solvability of the equation

$$\Box g = f$$

in the class of functions g with compact support, it is sufficient (and necessary) that $Mf(H(\omega, p)) = 0$ for all $\omega, q(\omega) = 0$ and all p. Here $\Box = \partial_0^2 - \Delta$ is the wave operator.

8.8 Decreasing solutions of Maxwell's system

Take Maxwell's system in vacuum instead of the wave operator

$$\begin{aligned} \nabla \times H - E_t' &= j, \\ (\nabla, E) &= \rho, \\ \nabla \times E + H_t' &= 0, \\ (\nabla, H) &= 0. \end{aligned} \tag{8.19}$$

We have used the Lorentz-Heaviside system of units to simplify the appearance. The right sides (j, ρ) should satisfy the consistency equation

$$(\nabla, j) + \rho_t' = 0. \tag{8.20}$$

We focus on the question, in which case does this system have a solution (E, H) with compact support? Note that there is no more than one such solution. Indeed, if $\rho = 0$, $j = 0$, any component of a solution fulfils the wave equation and vanishes everywhere if it has a compact support. For existence of the solution with compact support it is necessary that current j and the charge ρ have compact support and fulfil the consistency equation (8.20). This equation is not, however, sufficient. Introduce the differential form in the space-time $J = \rho \mathrm{d}t + j_1 \mathrm{d}x_1 + j_2 \mathrm{d}x_2 + j_3 \mathrm{d}x_3$.

Theorem 8.15 *The conditions*

$$R(\mathrm{d}J)(\omega, p) = 0, \ \omega = (\xi, \tau), \ |\xi|^2 - \tau^2 = 0, \ p \in \mathbb{R} \tag{8.21}$$

are necessary for the system (8.19) *to have a solution* (E, H) *with compact support in space-time. These conditions together with* (8.20) *are sufficient for existence of a solution with compact support.*

Noether operators. Let $A \doteq \mathbb{C}[\xi_1, \ldots, \xi_n]$ for some n and M be an A-module M of finite type. Let $\mathfrak{p}_1, \ldots, \mathfrak{p}_q$ be the set of all prime ideals in A associated to M ([54]). A $\mathbb{C}$-linear mapping $\delta : A^s \to \sum_1^q [A/\mathfrak{p}_j]^{k_j}$ is called a *Noether operator* for M, if k_j are some natural numbers, δ is a differential operator, $N \doteq \operatorname{Ker} \delta$ is a submodule of A^s and there is an isomorphism $M \simeq A^s/N$.

Example 6. Let $n = 1$, $M = A/\left(\xi^k\right)$. Then the maximal ideal $\mathfrak{m}$ is the only ideal associated to M. The operator $\delta : A \to [A/\mathfrak{m}]^k$, $a \mapsto \left(a(0), a'(0), \ldots, a^{(k-1)}(0)\right)$ is a Noether operator for M.

The following result was proved in [66]:

Theorem 8.16 *Let*

$$A^t \xrightarrow{p} A^s \xrightarrow{\delta} \sum_1^q [A/\mathfrak{p}_j]^{k_j}$$

be an exact sequence of vector spaces where p *is an* A*-morphism and* δ *is a differential operator. Then* δ *is a Noether operator for the module* $M = \operatorname{Cok} p$ *and for an arbitrary convex domain* $W \subset V$ *the sequence of vector spaces*

$$D^t(W) \xrightarrow{p} D^s(W) \xrightarrow{\delta} \sum_1^q [D(W)/\mathfrak{p}_j D(W)]^{k_j}$$

is also exact. The same is true if one replaces $D(W)$ *by the space* $E'(W)$ *of distributions with compact supports.*

Lemma 8.17 *Let* (K, d) *be the Koszul complex over the polynomial algebra* $A = \mathbb{C}[z_1, \ldots, z_n]$ *with the generators* $z_1, \ldots, z_n$*. Consider the mapping* $D : K \to K \oplus K$ *by* $a \mapsto (\mathrm{d}a, \mathrm{d}^* a)$*. The morphism*

$$N : (b, c) \mapsto \left\{ \begin{matrix} \mathrm{d}^* b + \mathrm{d}c \mod z^2 \\ \mathrm{d}b \\ \mathrm{d}^* c \end{matrix} \right\}$$

is a Noether operator for D*.*

◀ We have $N D a = (\mathrm{d}^*\mathrm{d} + \mathrm{d}\mathrm{d}^*) a = z^2 a$. Vice versa, let $N(b, c) = 0$. Then $\mathrm{d}^* b + \mathrm{d}c = z^2 e$ for some $e \in K$. Set $b_0 = \mathrm{d}e, c_0 = \mathrm{d}^* e$. We have then $N(b_0, c_0) = z^2 e$, hence, $\mathrm{d}^* b_1 + \mathrm{d}c_1 = 0$ for $b_1 = b - b_0, c_1 = c - c_0$. We have $z^2 b_1 = \mathrm{d}^*\mathrm{d}b_1 + \mathrm{d}\mathrm{d}^* b_1 = -\mathrm{d}\mathrm{d}c_1 = 0$ because $\mathrm{d}b_1 = 0$. This implies $b_1 = 0$ and similarly $c_1 = 0$. ▶

◀ Proof of Theorem. Write Maxwell's system in the relativistic form

$$\mathrm{d}^* F = J, \ \mathrm{d}F = I \tag{8.22}$$

where $F \doteq \sum E_i \mathrm{d}t \wedge \mathrm{d}x_i + \sum F_i \mathrm{d}x_j \wedge \mathrm{d}x_k$ and d^* means the adjoint operator with respect to Minkowski's metric $g_{ij} = \varepsilon_i \delta_{ij}$, $\varepsilon_0 = 1, \varepsilon_1 = \varepsilon_2 = \varepsilon_3 = -1$; I is a differential form of degree 3 (it vanishes in the standard form of Maxwell's system). Applying Theorem 8.16 to the Noether operator N, yields that the conditions

$$\mathrm{d}^* J = 0, \ \mathrm{d}I = 0, \ \mathrm{d}J + \mathrm{d}^* I = 0, \ \mathrm{mod}\, \square$$

are necessary and sufficient for solvability of (8.22) in the class of smooth forms F with compact supports in V or currents with compact supports. The first condition $\mathrm{d}^* J = 0$ coincides with (8.20). If $I = 0$, the third condition is equivalent to $F(\mathrm{d}J)(\omega) = 0$ as ω belongs to the cone $|\xi|^2 - \tau^2 = 0$. Applying the inverse Fourier transform along generators of this cone we come to (8.20). ▶

8.9 Symmetric differential forms

Let $\Sigma^k(\mathbf{V})$ be the space of smooth symmetric differential forms in a vector space $\mathbf{V}$ of dimension $n > 1$. Choose a coordinate system $x_1, \ldots, x_n$ in $\mathbf{V}$; a form $f \in \Sigma^k$ can be written as

$$f = \sum f_{i_1,\ldots,i_k}(x)\, \mathrm{d}x_{i_1} \ldots \mathrm{d}x_{i_k},$$

where $f_{i_1,\ldots,i_k}$ is a function $\mathbf{V}$ that is symmetric with respect to the indices; the product $\mathrm{d}x_{i_1} \ldots \mathrm{d}x_{i_k}$ is symmetric too. The number of independent components of f is equal to $\binom{n+k-1}{k}$. Define the symmetric differential

$$\begin{aligned} Df &= \sum \mathrm{d}f_{i_1,\ldots,i_k}(x)\, \mathrm{d}x_{i_1} \ldots \mathrm{d}x_{i_k} \\ &= \sum_{j_1,\ldots,j_{k+1}} \ \sum_{\pi(i_1,\ldots,i_k,j) = j_1,\ldots,j_{k+1}} \frac{\partial f_{i_1,\ldots,i_k}}{\partial x_j} \mathrm{d}x_{j_1} \ldots \mathrm{d}x_{j_{k+1}}, \end{aligned}$$

where the interior sum is taken over all substitutions π of $k+1$ elements. It defines a linear operator $D : \Sigma^k \to \Sigma^{k+1}$. Let Σ_0^k be the space of smooth symmetric forms with compact support. The image of the operator $D : \Sigma_0^k \to \Sigma_0^{k+1}$ can be described in terms of the flat integral transform in the Euclidean space, see [90].

Consider the case $k = 1$ for simplicity. Then Σ^2 is the space of sections of the bundle $S^2(\Omega)$ which is the symmetric square of the bundle $S^1(\Omega)$ of 1-forms in $\mathbf{V}$. Let Q^2 be the space of sections of $S^2(\Omega) \otimes S^2(\Omega)$ in $\mathbf{V}$. The Saint-Vennan operator is defined as

$$W : \Sigma^2 \to Q^2; \ g \mapsto Wg = \sum \sum (Wg)_{ij,kl}\, \mathrm{d}x_i \mathrm{d}x_j \otimes \mathrm{d}x_k \mathrm{d}x_l,$$

$$(Wg)_{ij,kl} = \frac{\partial^2 g_{ik}}{\partial x_j \partial x_l} - \frac{\partial^2 g_{il}}{\partial x_j \partial x_k} - \frac{\partial^2 g_{jk}}{\partial x_i \partial x_l} + \frac{\partial^2 g_{jl}}{\partial x_i \partial x_k},$$

in particular, for $i = k,\ j = l$

$$(Wg)_{ij,ij} = \frac{\partial^2 g_{ii}}{\partial x_j^2} - 2\frac{\partial^2 g_{ij}}{\partial x_i \partial x_j} + \frac{\partial^2 g_{jj}}{\partial x_i^2}.$$

It is easy to check that $WD = 0$.

Theorem 8.18 *The sequence*

$$\Sigma_0^1 \stackrel{D}{\to} \Sigma_0^2 \stackrel{W}{\to} \Sigma_0^{2,2}, \tag{8.23}$$

is exact. Moreover, if $Wg = 0$ for a form g with compact support, there exists a form f such that $Df = g$ and $\operatorname{supp} f \subset H(\operatorname{supp} g)$.

◀ Let $A = \mathbb{C}[z_1, \dots, z_n]$ be the algebra of polynomials in $\mathbf{V}^*$. To prove exactness of (8.23), we check exactness of the sequence of symbols of (8.23):

$$A^n \stackrel{\mathrm{d}}{\to} A^N \stackrel{w}{\to} A^M,$$

where $N = \binom{n+1}{2}$, $M = \binom{n^2+1}{2}$,

$$(\mathrm{d}a)_{ij} = z_i a_j + z_j a_i,$$
$$(wb)_{ij,kl} = z_j z_l b_{ik} - z_j z_k b_{il} - z_i z_l b_{jk} + z_i z_k b_{jl}.$$

Then exactness of (8.23) follows by Theorem 8.16. It is sufficient to check that the equations $(wb)_{ij,ij} = 0$ imply that $b = \mathrm{d}a$ for some $a \in A^n$. These equations are equivalent to

$$z_i^2 b_{kk} - 2z_i z_k b_{ik} + z_k^2 b_{ii} = 0$$

for arbitrary $k \neq i$. The first two terms are multiples of z_i which yields that so is b_{ii}, that is $b_{ii} = 2z_i a_i$ for some uniquely defined polynomial a_i. We have also $b_k = 2z_k a_k$ for some $a_k \in A$. Cancelling the factor $2z_i z_k$, we get the equation $b_{ik} = z_k a_i + z_i a_k$ which completes the proof. ▶

Remark. We have used only the "diagonal" part of the Saint-Vennan operator: $(Wg)_{ii,kk} = 0$.

Proposition 8.19 *The equation $wb = 0$ is equivalent to the condition*

$$\sum b_{ij}(\eta)\,\xi_i\xi_j = 0 \textit{ for arbitrary points } \xi, \eta \in \mathbf{V}^* \textit{such that } (\eta, \xi) = 0.$$

◀ Take an arbitrary point $\eta \in \mathbf{V}^*$ and set $\xi = \eta_k e_l - \eta_l e_k$ where $e_1, \dots, e_n$ is a basis in $\mathbf{V}^*$. Then

$$\sum b_{ij}(\eta)\,\xi_i\xi_j = b_{ll}\eta_k\eta_k + b_{kk}\eta_l\eta_l - 2b_{kl}\eta_k\eta_l = (wb)_{kl,kl}\,.$$

The condition $wb = 0$ implies that the right side vanishes. The vectors $\xi^{(kl)} = \eta_k e_l - \eta_l e_k$ are dense in the plane orthogonal to η, hence the left side vanishes in this plane. Vice versa, if the left side vanishes for $(\xi, \eta) = 0$, it vanishes for $\xi = \xi^{(kl)}$ which yields $(wb)_{kl,kl} = 0$. According to the previous proof, this implies that $b \in \operatorname{Im} \mathrm{d}$ and $wb = 0$. ▶

Theorem 8.20 *For a symmetric form f with compact support in $\mathbf{V}$, the integral equation*

$$\sum \int f_{ij}(x + t\xi)\,\xi_i \xi_j \mathrm{d}t = 0 \tag{8.24}$$

is a necessary and sufficient condition for existence of a form g with compact support fulfilling $f = Dg$. Moreover, $\operatorname{supp} g \subset H(\operatorname{supp} f)$.

◀ The condition (8.24) is equivalent to the condition

$$\sum \hat{f}_{ij}(\eta)\,\xi_i \xi_j = 0 \text{ as } (\xi, \eta) = 0$$

for the Fourier transforms $\hat{f}_{ij}$ of the functions f_{ij}. According to Proposition 8.19 the latter is equivalent to $w\hat{f} = 0$. By Theorem 8.18 the last equation holds if and only if the equation $\mathrm{d}\hat{g} = \hat{f}$ has a solution $g \in \Sigma_0^1$. By the inverse Fourier transform we get $Dg = f$. ▶

Theorem 8.21 *The sequence*

$$\Sigma^1 \xrightarrow{D} \Sigma^2 \xrightarrow{W} \Sigma^{2,2}$$

is exact.

◀ We need only to check exactness of the dual sequence of symbols:

$$A^M \xrightarrow{w^*} A^N \xrightarrow{\mathrm{d}^*} A^n,$$

where

$$(\mathrm{d}^* b)_i = \sum z_j b_{ij},\ (w^* c)_{ij} = \sum z_k z_l \left(c_{ik,jl} - c_{il,kj} - c_{kj,il} + c_{lj,ki}\right).$$

We need to prove that the equation $\mathrm{d}^* b = 0$ implies that $b = w^* c$ for some polynomial vector c. Consider, first, the case $n = 2$; the equation $\mathrm{d}^* b = 0$ reads

$$z_1 b_{11} + z_2 b_{12} = 0,\ z_1 b_{21} + z_2 b_{22} = 0,\ b_{21} = b_{12}.$$

By the first and the second equations we have $b_{11} = z_2 b_1, b_{12} = -z_1 b_1$, $b_{21} = z_2 b_2$, $b_{22} = -z_1 b_2$ for some polynomials b_1, b_2. The third one yields $b_1 = z_2 b, b_2 = -z_1 b$ for a polynomial b. Whence $b_{11} = z_2^2 b, b_{12} = -z_1 z_2 b, b_{22} = z_1^2 b$. This means that $b = w^* c$ where $c_{11,22} = b/2$, $c_{12,12} = 0$. In the general case we use induction on $n > 2$. The equation $\mathrm{d}^* b = 0$ implies $\mathrm{d}'^* b' = 0$ where $b'_{ij} \doteq b_{ij} \pmod{z_n}$, $i, j < n$ and d' denotes the operator d in the space $\mathbf{U} = \{z \in (V)^* z_n = 0\}$. By inductive assumption we have $b' = w'^* c'$ for some vector c' whose components are $A' \doteq \mathbb{C}[z_1, \ldots, z_{n-1}]$ and w' is the symbol of the Saint-Vennan operator in $\mathbf{U}$. Applying the standard embedding $A' \to A$ we consider c' as an element of A^M;

set $\beta \doteq b - w^* c'$. We have $\beta_{ij} = z_n \gamma_{ij}$ for some polynomials $\gamma_{ij}, i, j < n$. The equation $\mathrm{d}^* \beta = 0$ implies

$$z_n \beta_{in} = -\sum_{j=1}^{n-1} z_j z_n \gamma_{ij},\ \beta_{in} = -\sum_1^{n-1} z_j \gamma_{ij},\ \sum_1^n z_i \beta_{in} = 0$$

and

$$z_n \beta_{nn} = -\sum_1^{n-1} z_i \beta_{in} = \sum_1^{n-1} z_i z_j \gamma_{ij}.$$

This equation implies that $\beta_{nn} = \sum z_i z_j c_{ij}$ for some polynomials c_{ij}. Set $c''_{nn,ij} = c_{ij}$ and $c''_{ij,kl} = 0$ otherwise. Consider the vector $b'' = \beta - w^* c''$. Now we have $b''_{nn} = 0$, whence

$$\sum_1^{n-1} z_i z_j \gamma_{ij} = 0,$$

which yields $\gamma_{ij} = \sum_{k=1}^{n-1} z_k d_{k,ij}$ for some polynomials $d_{k,ij} = d_{k,ji} = -d_{i,kj}$ and $b''_{in} = -\sum z_j z_k d_{k,ij}$. Set $c'''_{kn,ij} = d_{k,ij}$ and $c'''_{ij,kl} = 0$ otherwise; and take the vector $b''' = b'' - w^* c'''$. Then we have $b'''_{in} = 0$ for all i and we can again apply the inductive assumption. ▶

Chapter 9

Notes

Chapter 1

Method of analytic continuation of distributions and Proposition 1.6 are due to M. Riesz [84].

See [88], [33] and [71] for basics of the distribution theory and [18] for the integration theory An elementary introduction to Fourier integrals and sampling theory is contained in [3], [21]. Proposition 1.22 was published in [77]

Chapter 2

Note some pioneering works on applied problems: [10], [8], [19]. The historical surveys can be found in [91], [15] and in [61]. The Radon transform and its generalizations are the focus of [15], [61], [43], [44].

To Section 2.3. Theorem 2.8 is due to Gindikin [34], see also [29] and the paper [96] for a numerical implementation. Uniqueness of reconstruction from integrals over closed geodesics was shown in [98].

To Section 2.4. Description of the range of the Radon transform is called the Helgason-Ludwig theorem, see [44].

To Section 2.5. Theorem 2.12 is a sharp form of the Lindgren-Rattey property of the Fourier series of the Radon transform in parallel and in fan-beam geometries, [72]. This property is exploited for the efficient reconstruction algorithms [61], [60]. Henkin and Shananin [45] gave a description of Radon images of positive measures supported by the coordinate "octant" $\mathbb{R}^n_+$.

To Section 2.6. The integral transform of differential forms is the mathematical background of the "vector tomography", see [4].

Chapter 3

H. Minkowski [58] proved that an even function f on the Euclidean 2-sphere is uniquely determined by its integrals along big circles. P. Funk [26] found an explicit reconstruction formula for f. He took the average of Mf along orbits of an actions of the rotation group on the sphere and showed that the average of Mf is related to the average of f by the Abel transform. J. Radon studied the similar problem for the Euclidean plane, see [82] (or [43]).

The method of factorable mapping has been suggested in [69] for projective transforms in Euclidean space and in [74] in the general setting. See also papers of Denisjuk [16], [17], where this method was applied to spaces of constant curvature. The reconstruction formulae for the family of curves in Example 6 were found by Cormack [11].

The results of Section 3.2 were published in [75].

To Section 3.4. The Funk transform was studied by Semyanistyi [89]. It was generalized by Helgason, [44] for symmetric spaces as the totally geodesic transform, see also the paper of Berenstein and Tarabusi [5]. Another generalization of the Minkowski-Funk transform was studied in [87]: Euler kernels instead of the Dirac density on big spheres.

The text of Section 3.5 follows the paper [28] up to some details.

To Section 3.6. A description of the wave front in terms of the Radon transform of a distribution was done in a similar form in the book of Guillemin and Sternberg [41]. The Funk-Radon transform can be considered as a particular case of Hörmander's theory of Fourier integral operators, [46]. Homogeneous canonical relations play the role of the classical Legendre transform in the general case.

To Section 3.7. Theorem 3.13 is, of course, known in several situations. In particular, in [32] Radon transforms of f_λ for quadratic functions ϕ are explicitly calculated for several representative cases, the dual function ψ is again quadratic. A general case is addressed in the paper [81]. The Radon images of more complicated singular functions were studied in [68].

To Section 3.8. A mathematical model of nonlinear artifacts in 2D-tomography was proposed in [67].

Chapter 4

A reconstruction from data of ray integrals with sources at infinity was done by Orlov [65]. The case of proper curve was considered by Kirillov [50] (uniqueness), Tuy [95], Finch [24], Grangeat [38] (reconstruction). A. Katsevich [49] has adapted Grangeat's method for numerical implementation.

The result of Section 4.4 was stated in [69]. The result of Section 4.5 is published in [76]. Theorem 4.12 is new.

To Section 4.7. The integral transforms of differential forms in an affine and projective space are the focus of the papers [28] and [29]. The line integral transform of differential 1-forms and the plane transform of 2-forms are mathematical models for tomography of vector fields, see [4]. A more general construction is the line transform of symmetric differential forms, see also Section 8.9. The theory and applications of this transform are contained in the book [90].

To Section 4.8. An inversion formula of the exponential ray transform was found by Tretiak and Metz [94]. Another approach was developed by Natterer [59],

[61]. We follow the method of Kuchment and Shneiberg [51]. The description of the range of the exponential transform was obtained in [1] for arbitrary dimension.

To Section 4.10. The first reconstruction formula for the general attenuated ray transform in 2D case was found by R. Novikov [63]. Natterer [62] simplified Novikov's original proof and found Theorem 4.24. Lemma 4.22 is taken from [61], Theorem 4.23 is obtained in [63], see also [64]. For the 3D case see [73].

To Section 4.11. Theorem 4.25 is due to Natterer [61].

Chapter 5

The method of the operator κ was proposed by Gelfand, Graev and Shapiro, [31], [30] for reconstructions from integrals over affine subspaces of even dimension. See also [29]. The inversion formula proposed in [30] is not quite satisfactory: since the integral operator RIR may diverge.

To Section 5.3 The results for odd dimensional case are new.

For Helgason's theorem see [43]. The general case was indicated in [28]. A complete proof was done by F. Richter [83]. The differential range condition was found by F. John [47] for the case $n = 3$, $d = 1$. The general case was studied in [83]. Further results concerning moment conditions and John-type equations can be found in [83] and [36]. Theorem 5.9 was published in [52]. The characterization of the range of the projective integral d-transform needs no conditions at infinity, [28]. The range is essentially equal to the space of solutions of the system of John's equations 5.8.

Duality theorems were published in [74].

Chapter 6

To Section 6.1. The completeness condition appeared in papers of Kirillov [50] and Tuy [95], see also [68]. The pseudodifferential operator $R^\sharp R$ was studied in a general situation by Guillemin and Sternberg [40]; it is shown in [53] that the composition $R^\sharp R$ is a classical singular integral operator. Application of the operator $R^\sharp$ to the line transform Rf in a plane yields a related function $\tilde{f}$, which has the same singularity. This operation is local in the sense that only lines through a given point are involved. This is the basic idea of "local tomography", see [22]. D. Popov [78] found a quasi inverse operator in the form $R^\sharp B$, where B is an elliptic PDO.

To Section 6.2. Problems with incomplete data were considered by Natterer [61], Louis [57], Quinto [79], [80]. Lent and Tuy [56] developed a method based on interpolation of bandlimited functions and a priori information. We follow the paper [74], which contains additional information.

To Section 6.4. Theorem 6.5 is due to Cormack in the case $n = 2$. The general case was obtained by Deans [14]. A non-uniqueness is shown by J.Boman in [7] for the integral transform with special smooth weight.

In Section 6.5 we follow to G.Beylkin [6].

Chapter 7

To Section 7.1. For applications of spherical transform to differential equations see [48], [13], to seismic tomography see Romanov [85], [86]; for applications to synthetic radar aperture see [42], to scattering problem [9] and in thermoacoustic tomography [25]. The problem of reconstruction from hemispherical integrals has been studied by several authors. Inversion methods for the case $n = 2$ were given in [85] and [42]. See [29] for the equivalent problem in the hyperbolic plane. A solution for $n = 3$ is given in [9]. Fawcett [23] found an explicit formula for the solution for an arbitrary n.

To Section 7.2. The contents is taken from [75]. The inversion from limited data is addressed also in [2] (local problem). In Sections 7.3, 7.4 we follow [17].

To Section 7.5. Theorem 4.10 is close to a formula published the paper of Finch, Patch and Rakesh [25].

To Section 7.6. The case of spheres through a point is considered in [12]. Gelfand and Graev studied the transform given by horocycle integrals in a homogeneous space. For the standard model of hyperbolic space horocycles are just the spheres tangent to the boundary of the ball. The reconstruction formulae are given in the book [32]. For further information see [44].

Gindikin [35], [29] and Goncharov [37] addressed the reconstruction problem for functions in $\mathbb{R}^n$ from data of integrals over spheres tangent to a boundary of a domain D. Their method is based on reduction to the Radon transform in a projective quadric. The class of "admissible" submanifolds in the manifold of spheres in the sense of [37] and also of the sense [29] essentially coincides with the class of characteristic hypersurfaces. Results of Sections 7.6, 7.7 and 7.8 are new. The singular Cauchy problem for the Euler-Poisson-Darboux equation was studied in [20]; see therein a survey of previous results.

Chapter 8

To Section 8.3: the description of the characteristic variety of the multiplicative system was published in [70]. Contents of Sections 8.4 and 8.5 is new. The results of Sections 8.7 and 8.8 are based on [66]. In Section 8.10, we deduce some results of Sharafutdinov's book [90].

Bibliography

[1] Aguilar, V., Kuchment, P., (1995). Range conditions for the multidimensional exponential X-ray transform *Inverse Problems.* 11: N5, 977–982.

[2] Andersson, L.E., (1988). On determination of a function from spherical averages *SIAM J Math Anal.* 19: 214-232.

[3] Arsac, J., (1961). Fourier transforms and the theory of distributions. Prentice Hall.

[4] Braun, H., Hauck, A., (1991). Tomographic reconstruction of vector fields *IEEE Trans Signal Process.* 39: 464–471.

[5] Berenstein, C.A., Tarabusi, E.C., (1993). Range of the k-dimensional Radon transform in real hyperbolic spaces *Forum Math.* 5: 603–616.

[6] Beylkin, G., (1984). The inversion problem and applications of the generalized Radon transform *Comm. Pure Appl. Math.* 37: N5, 579–599.

[7] Boman, J., (1993). An example of nonuniqueness for a generalized Radon transform *J. Anal. Math.* 61: 395 401.

[8] Bracewell, R.N., Riddle, A.C., (1967). Inversion of fan beam scan in radio astronomy *Astrophys. J.* 150: 427–434.

[9] Cohen, J., Bleistein, N., (1979). Velocity inversion procedure for acoustic waves *Geophysics* 44: 1077–1085.

[10] Cormack, A.M., (1963). Representation of a function by its line integrals, with some radiological applications I. *J. Appl. Phys.* 34: 2722–2727.

[11] Cormack, A.M., (1981, 1982). The Radon transform on a family of curves in the plane I. *Proc. Amer. Math. Soc.* 83: 325–330; II *Ibid* 86: 293–298.

[12] Cormack, A.M., Quinto, E.T., (1980). A Radon transform on spheres through the origin in $\mathbb{R}^n$ and applications to the Darboux equation. *Transl. Amer. Math. Soc.* 260: 575–581.

[13] Courant, R., Hilbert, D., (1962). *Methods of Mathematical Physics II.* New York-London: Interscience Publ.

[14] Deans, S.R., (1979). Gegenbauer polynomials via the Radon transform. *SIAM J. Math. Anal.* 10: 577–585.

[15] Deans, S.R., (1983). Tha Radon transform and some of its applications. J. Wiley.

[16] Denisjuk, A.S., (1997). Inversion of the X-ray transform in spaces of constant curvature. *Siberian Math. J.* 38: N5, 1058–1070.

[17] Denisjuk, A.S., (1999). Integral geometry on the family of semi-spheres. *Frac. Calc. Appl. Anal.* 2: N1, 31–46.

[18] De Rham, G., (1955). *Variétés Différentiables.* Paris: Hermann.

[19] De Rozier, D., Klug, A., (1968). Reconstruction of three-dimensional structures from electron micrographs. *Nature.* 217: 130–138.

[20] Diaz, J.B., Weinberger, H.F., (1953). A solution of the singular initial value problem for Euler-Poisson-Darboux equation. *Proc. Amer. Math. Soc.* 4: 703–715.

[21] Epstein, Ch.L., (2003). *Introduction to the Mathematics of Medical Imaging.* NJ: Rearson, Prentice Hall.

[22] Faridani, A., Finch, D.V., Ritman, E.L., Smith, K.T., (1997). Local tomography II. *SIAM J. Appl. Math.* 57: N4, 1095–1127.

[23] Fawcett, J.A., (1985). Inversion of n-dimensional spherical averages. *SIAM J. Appl. Math.* 45: 336–341.

[24] Finch, D., (1985). Cone beam reconstruction with sources on a curve. *SIAM J. Appl. Math.* 45: N4, 665–673.

[25] Finch, D., Patch, S.K., Rakesh, (2004). Determining a function from its mean values over a family of spheres. *SIAM J. Math. Anal.* 35: 1213–1240.

[26] Funk, P., (1916). Über eine geometrische Anwendung der Abelschen Integralgleichung. *Math. Ann.* 77: 129–135.

[27] Gelfand, I.M., Gindikin, S.G., (1977). Nonlocal inversion formulas in real integral geometry. *Funct. Anal. Appl.* 11: N3, 173–179.

[28] Gelfand, I.M., Gindikin, S.G., Graev, M.I., (1980). Integral geometry in affine and projective spaces. *J. Sov. Math.* 18: 39–167.

[29] Gelfand, I.M., Gindikin, S.G., Graev, M.I., (2003). *Selected topics in Integral Geometry.* Transl. of Math. Monographs 220. Providence RI: Amer. Math. Soc.

[30] Gelfand, I.M., Graev, M.I., (1991). Crofton function and an inversion formula in real integral geometry. *Funct. Anal. Appl.* 25: N1, 1–6.

[31] Gelfand, I.M., Graev, M.I., Shapiro, Z.Ya., (1967). Integral geometry on k-dimensional planes. *Funkt Anal Appl.* 1: N1, 15–31.

[32] Gelfand, I.M., Graev, M.I., Vilenkin, N.Ya., (1966). *Integral Geometry and Representation Theory.* New York-London: Academic Press.

[33] Gelfand, I.M., Shilov, G.E., (1964). *Generalized functions Vol 1.* New York-London: Academic Press.

[34] Gindikin, S.G., (1966). A universal integral formula for multicircular regions in n-dimensional complex space. *Siberian Math. J.* 7: 708–712.

[35] Gindikin, S.G., (1995). Integral geometry on real quadrics. *Amer. Math. Soc. Transl. Ser. 2, 169.* Providence RI: Amer. Math. Soc., 23–31.

[36] Gonzalez, F.B., (1990). Invariant differential operators and the range of the Radon D-plane transform. *Math. Ann.* 287: N4, 627–635.

[37] Goncharov, A.B., (1997). Differential equations and integral geometry. *Adv. Math.* 131: N2, 279–343.

[38] Grangeat, P., (1991). Mathematical framework of cone-beam 3D-reconstruction via the first derivative of the Radon transform. *Lecture Notes in Math. 1497* Springer, 66–97.

[39] Guillemin, V. (1985). On some results of Gelfand in integral geometry. Pseudodifferential operators and applications, Providence RI: Amer. Math. Soc., 149–155.

[40] Guillemin, V., Sternberg, S., (1979). Some problems in integral geometry and some related problems in microlocal analysis. *Amer. J. Math.* 101: N4, 915–955.

[41] Guillemin, V., Sternberg, S., (1997). *Geometrical asymptotics.* Mathematical Surveys 14, Providence RI: Amer. Math. Soc.

[42] Hellsten, H., Andersson, L.E., (1987). An inverse method for the processing of synthetic aperture radar data. *Inverse Problems* 3: 111–124.

[43] Helgason, S., (1994). *Geometric analysis in symmetric spaces.* Providence RI: Amer. Math. Soc.

[44] Helgason, S., (1999). *The Radon transform, Second edition.* Boston: Birkhäuser.

[45] Henkin, G.M., Shananin, A.A., (1990). Bernstein theorems and Radon transform. Application to the theory of productive functions. *Mathematical problems of tomography Transl. Math. Monographs 81.* Providence RI: Amer. Math. Soc, 189–223.

[46] Hörmander, L., (1971). Fourier integral operators I. *Acta Math.* 127: N1-2, 79–183.

[47] John, F., (1938). The ultrahyperbolic differential equation with four independent variables. *Duke Math. J.* 4: 300–322.

[48] John, F., (1955). *Plane waves and spherical means applied to partial differential equations.* Interscience.

[49] Katsevich, A., (2002). Theoretically exact filtered back projection-type inversion algorithm for spiral CT. *SIAM J. Appl. Math.* 62: 2012–2026.

[50] Kirillov, A.A., (1961). On a problem of I.M. Gelfand. *Soviet Math. Dokl.* 2: 268–269.

[51] Kuchment, P., Shneiberg, I., (1994). Some inversion formulas in the single photon emission computed tomography *Applicable Analysis.* 53: 221–231.

[52] Kurusa, A., (1991). A characterization of the Radon transform's range by a system of PDE's. *J. Math. Anal. Appl.* 161: 218–226.

[53] Lavrent'ev, M.M., Romanov, V.G., Shishatskii, S.P., (1986). *Ill-posed problems of mathematical physics and analysis.* Transl. Math. Monographs. 64 Providence RI: Amer. Math. Soc.

[54] Lang, S., (1965). *Algebra.* Mass: Addison-Wesley, Reading.

[55] Lax, P.D., Phillips, R.S., (1967). *Scattering theory.* New York: Academic Press.

[56] Lent, A., Tuy, H., (1981). An iterative nethod for the extrapolation of band-limited functions. *J. Math. Anal. Appl.* 83: N2, 554–565.

[57] Louis, A.K., (1986). Incomplete data problems in X-ray computerized tomography I; Singular value decomposition of the limited angle problem. *Numer Math.* 48: 251–262.

[58] Minkowski, H., (1904). On bodies of constant width [in Russian]. *Mat. Sbornik.* 25: 505–508; German translation in: *Gesammelte Abhandlungen 2* Leipzig: Teubner 1911, 277–279.

[59] Natterer, F., (1979). On the inversion of the attenuated Radon transform. *Numer. Math.* 32: 431–438.

[60] Natterer, F., (1993). Sampling in Fan Beam Tomography. *SIAM J. Appl. Math.* 53: N2, 358–380.

[61] Natterer, F., (1986). *The mathematics of computerized tomography.* Stuttgart: Teubner.

[62] Natterer, F., Wübbeling, F., (2001). *Mathematical Methods in Image Reconstruction.* Philadelphia: SIAM.

[63] Novikov, R.G., (2002). An inversion formula for the attenuated X-ray transform. *Arkiv för matematik.* 40: N1, 145–167.

[64] Novikov, R.G., (2002). On the range characterization for the two-dimensional attenuated X-ray transformation. *Inverse Problems* 18: 677–700.

[65] Orlov, S., (1975). Theory of three-dimensional reconstruction II. The recovery operator *Soviet Phys. Cristallogr.* 20: N4, 429–433.

[66] Palamodov, V.P., (1967). *Linear differential operators with constant coefficients.* Moscow: Nauka; Springer (1970).

[67] Palamodov, V.P., (1986). Nonlinear artifacts in tomography. *Soviet Phys. Dokl.* 31(11): 888–890.

[68] Palamodov, V.P., (1990). Some singular problems in tomography. *Mathematical Problems of Tomography. Transl. Math. Monographs, 81.* Providence RI: Amer. Math. Soc., 123–140.

[69] Palamodov, V.P., (1991). Inversion formula for the three-dimensional ray transform, In: *Lecture Notes in Mathematics, 1497* Springer, 53–62.

[70] Palamodov, V.P., (1994). Radon transformation on real algebraic varieties, In: *Proceedings of the Conference 75 years of Radon transform.* Cambridge MA: International Press, 252–262.

[71] Palamodov, V.P., (1995). Distributions and Harmonic Analysis, In: *Encyclopaedia of Mathematical Science, 72* Springer 1–127.

[72] Palamodov, V.P., (1995). Localization of harmonic spectrum of Radon transform *Inverse Problems.* 11: N5, 1025–1030.

[73] Palamodov, V.P., (1996). An inversion method for an attenuated X-ray transform. *Inverse Problems.* 12: N5, 717–729.

[74] Palamodov, V.P., (1998). Reconstruction from line integrals in spaces of constant curvature. *Math. Nachr.* 196: 167–188.

[75] Palamodov, V.P., (2000), Reconstruction from limited data of arc means. *J. of Fourier Anal. Appl.* 6: N1, 25–42.

[76] Palamodov, V.P., (2004). Reconstruction from ray integrals with sources on a curve. *Inverse Problems.* 20: N1, 239–242.

[77] Palamodov, V.P., Denisjuk, A.S., (1988). Inversion de la transformation de Radon d'après des données incomplètes. *C. R. Acad. Scie. Paris* 307: 181–183.

[78] Popov, D.A., (2001). The generalized Radon transform on the plane, the inverse transform and the Cavalieri conditions. *Funct. Anal. Appl.* 35: N4, 270–283.

[79] Quinto, E.T., (1993). Singularities of the X-ray transform and limited data tomography in $\mathbb{R}^2$ and $\mathbb{R}^3$. *SIAM J. Math. Anal.* 24: N5, 1215–1225.

[80] Quinto, E.T. (1998). Exterior and limited-angle tomography in non-destructive evaluation. *Inverse Problems.* 14: 339–353.

[81] Ramm, A., Zaslavsky, A., (1993). Singularities of the Radon transform *Bull. Amer. Math. Soc.* 28: 109–115.

[82] Radon, J., (1917). Über die Bestimmung von Funktionen durch ihre Integralwerte längs gewisser Mannigfaltigkeiten. *Berichte Sächsische Akademie der Wissenschaften, Math.-Phys. Kl.* 69: 262–267.

[83] Richter, F., (1986). On the k-dimensional Radon transform of rapidly decreasing functions, In: *Lecture Notes in Math., 1209* Springer, 243–258.

[84] Riesz, M., (1949). L'intégral de Riemann-Liouville et le problèleme de Cauchy *Acta Math.* 81: 1–223.

[85] Romanov, V.G., (1969). *Integral Geometry and Inverse Problems for Hyperbolic Equations.* Springer.

[86] Romanov, V.G., (1969). A problem of integral geometry an a linearized inverse problem for a hyperbolic equation. *Siberian Math. J.* 10(6): 1011–1018.

[87] Rubin, B., (2000). Generalized Minkowski-Funk transform and small denominators on the sphere. *Frac. Calc. Appl. Anal.* 3: N2, 177–203.

[88] Schwartz, L., (1966). *Théorie des distributions.* Paris: Hermann.

[89] Semyanistyi, V.I., (1961). Homogeneous functions and some problems of integral geometry in spaces of constant curvature. *Soviet Math. Dokl.* 2: 59–62.

[90] Sharafutdinov, V.A., (1994). *Integral Geometry of Tensor Fields.* Utrecht: VSP.

[91] Shepp, L.A., Kruskal, J.B., (1978). Computerized tomography: the new medical X-ray technology. *Amer. Math. Monthly.* 85: 420–439.

[92] Shubin, M.A., (1987). *Pseudodifferential operators and spectral theory.* Springer.

[93] Sneddon, I.H., (1972). *The use of integral transforms.* New York: McGraw-Hill.

[94] Tretiak, O., Metz, C., (1980). The exponential Radon transform. *SIAM J. Appl. Math.* 39: 341–354.

[95] Tuy, H.K., (1983). An inversion formula for cone-beams reconstruction. *SIAM J. Appl. Math.* 43: 546–552.

[96] Vvedenskaya, N.D., Gindikin, S.G., (1990). Discrete Radon transform and image reconstruction, In: *Mathematical Problems of Tomography, Transl. of Math. Monographs, 81* Providence RI: Amer. Math. Soc., 141–188.

[97] Wertgeim, L., (1992). Personal communication.

[98] Wolf, J.H., (1967). *Spaces of constant curvature.* McGraw-Hill.

Index